Student Study Guide

E S S E N T I A L S   O F
Human Anatomy and Physiology
Third Edition

Student Study Guide

E S S E N T I A L S O F
Human Anatomy and Physiology

Third Edition

John W. Hole, Jr.

Nancy A. Sickles Corbett
Director of Education
Zurbrugg Memorial Hospitals

wcb
Wm. C. Brown Publishers
Dubuque, Iowa

Cover illustration © William B. Westwoood

The credits section for this book begins on page 161, and is
considered an extension of the copyright page.

ISBN 0-697-01179-8

Printed in the United States of America by Wm. C. Brown Publishers
2460 Kerper Boulevard, Dubuque, IA 52001

10 9 8 7 6 5 4 3 2 1

Contents

To the Student

A study guide attempts to do as the term implies; that is, guide your study so that your learning efforts are most efficient.

This study guide is built on several beliefs: (1) learning occurs best when the learner is active rather than passive; (2) learning is easiest when the material is organized in simple units; and (3) the learner can best evaluate what he or she knows well, is unsure of, and does not know.

The study guide chapters correspond to the chapters in the text, *Essentials of Human Anatomy and Physiology,* Third Edition by John W. Hole, Jr., Wm. C. Brown Publishers, 1989. The elements of the study guide chapters and their purposes are described below.

1. *Overview.* The learning objectives at the beginning of each chapter in the text are arranged in groups according to broad, general concepts presented in the chapter. The overview also contains a purpose statement that offers a rationale for studying the chapter.
2. *Chapter Objectives.* The chapter objectives from the text are listed.
3. *Focus Question.* The focus question helps you focus your study of each chapter.
4. *Mastery Test.* The mastery test, taken before reading the chapter, is designed to help you identify:
   a. the concepts you already know,
   b. those concepts you need to clarify, and
   c. those concepts you do not know.

   If you are using a study guide for the first time, you may be unfamiliar with this type of testing. It is important for you to realize that this test is *for your information.* Its purpose is to help you learn where to concentrate your learning efforts; therefore, it is best not to guess at any answers.
5. *Study Activities.* A variety of study activities help facilitate the study of the principal ideas of each chapter.

The study activities should be done after you have read the chapter carefully, concentrating on those areas that the mastery test indicates you do not know.

The first activity in each unit is a vocabulary exercise, concentrating on word parts appropriate to each chapter. You are asked to define these as you understand them and then compare your definitions with those in the chapter. You may find it helpful to define terms orally and in writing. If you have a tape recorder, you may use it as a study device.

After the vocabulary exercise, you may be asked to describe a process, label a diagram, fill in a chart, or observe the function of a body part in yourself or in a partner. (This partner may be a classmate or a cooperative family member.) These are written activities, but you may also find it helpful to repeat them orally.

After you complete the study activities, retake the mastery test. A comparison of the two scores will indicate the progress you have made. You may also wish to set a learning goal for yourself, such as a score of 70%, 80%, or 90% on the mastery test after completing your study of a chapter. If you have not attained your goal, the mastery test results can show where you need additional study.

The answers to the mastery test are at the end of the study guide. You can compare your responses to the review activities by referring to the appropriate page numbers in the text. Each major section in the study guide is identified by a Roman numeral and the title of the corresponding section in the text. The activities in the study guide are lettered, and the corresponding pages in the text are noted after the activity.

You are responsible for your own learning; no teacher can assume that responsibility. A study guide can help you direct your study more efficiently, but only you can control how well and how completely you use the guide.

1 An Introduction to Human Anatomy and Physiology

Overview

This chapter begins the study of anatomy and physiology by defining the disciplines (objective 1) and explaining the characteristics and needs that are common to all living things (objectives 2 and 3). It introduces a basic mechanism necessary to maintain life (objectives 4 and 5), as well as the relationship of increasingly complex levels of organization in humans (objective 6). The study of levels of organization continues with the identification of body cavities and the organs to be found within each cavity (objectives 7 and 8). The membranes associated with the abdominopelvic and thoracic cavities are described (objective 9). The functions of the various organ systems as well as the organs associated with each system are described (objectives 10 and 11). Finally, the language used to describe relative positions of body parts, body sections, and body regions is presented (objective 12).

This chapter defines the characteristics and needs common to all living things and the manner in which the human body is organized to accomplish life processes. The language peculiar to anatomy and physiology is also introduced.

Chapter Objectives

After you have studied this chapter, you should be able to

1. Define *anatomy* and *physiology* and explain how they are related.
2. List and describe the major characteristics of life.
3. List and describe the major needs of organisms.
4. Define *homeostasis* and explain its importance to survival.
5. Describe a homeostatic mechanism.
6. Explain what is meant by levels of organization.
7. Describe the location of the major body cavities.
8. List the organs located in each of the body cavities.
9. Name the membranes associated with the thoracic and abdominopelvic cavities.
10. Name the major organ systems of the body and list the organs associated with each.
11. Describe the general functions of each organ system.
12. Properly use the terms that describe relative positions, body sections, and body regions.

Focus Question

How is the human body organized to accomplish those tasks that are essential to maintain life?

Mastery Test

Now, take the mastery test. Do not guess. As soon as you complete the test, correct it. Note your successes and failures so that you can read the chapter to meet your learning needs.

1. Study of the human body began with earliest humans because
   a. our early ancestors were curious about the world around them.
   b. they were as interested in their body parts and their functions as we are today.
   c. of their concern with illness and injury.

2. Which of the following factors set the stage for the development of modern science?
   a. a belief that spirits or gods controlled sickness and health
   b. the growing experience of medicine men as they treated the sick with herbs and potions
   c. the belief that natural processes were caused by forces that could be understood
   d. the ability to ask questions and record the answers

3. What languages form the basis of the language of anatomy and physiology?

4. The branch of science that deals with the structure of body parts is _____.

5. The branch of science that studies how body parts function is _____.

6. The function of a part is (always, sometimes, never) related to its structure.

7. List those characteristics that humans share with other organisms.

   a.   f.

   b.   g.

   c.   h.

   d.   i.

   e.   j.

8. The most abundant compound in the human body is _____.

9. Food is used as a(n) _____ source to build new _____ _____ and to participate in chemical reactions.

10. Oxygen is used to release _____.

11. An increase in temperature _____ the rate of chemical reactions.

12. The action of the heart creates _____ pressure in the blood vessels.

13. Homeostasis means

    a. maintenance of a stable internal environment.   c. preventing any change in the organism.

    b. integrating the functions of the various organ systems.

14. List the levels of organization of the body in order of increasing complexity, beginning with the cell.

    _____

    _____

15. The portion of the body that contains the head, neck, and trunk is called the _____ _____.

16. The arms and legs are called the _____ portion.

17. The two major cavities of the axial portion of the body are the _____ cavity and the _____ cavity.

18. The inferior boundary of the thoracic cavity is the _____.

19. The heart, esophagus, trachea, and thymus gland are located in the _____ of the thoracic cavity.

20. The pelvic cavity is

    a. the lower third of the abdominopelvic cavity.

    b. the portion of the abdomen that contains the reproductive organs.

    c. the portion of the abdomen surrounded by the bones of the pelvis.

21. The visceral and parietal pleural membranes secrete a serous fluid into a potential space called the

    _____ _____.

22. The heart is covered by the _____ membranes.

23. The peritoneal membranes are located in the ___abdominal___ cavity.

24. An organ is one organizational step higher than

    a. a system.   c. organelles.

    b. a cell.   d. tissue.

2

25. Match the systems listed in the first column with the functions listed in the second column.

   ____ a. nervous system
   ____ b. muscular system
   ____ c. circulatory system
   ____ d. respiratory system
   ____ e. skeletal system
   ____ f. digestive system
   ____ g. lymphatic system
   ____ h. endocrine system
   ____ i. urinary system
   ____ j. reproductive system

   1. reproduction
   2. processing and transporting
   3. integration and coordination
   4. support and movement

26. Which of the following positions of body parts is(are) in *anatomic* position?
   a. palms of hands turned toward sides of body
   b. standing erect
   c. arms at side
   d. face toward left shoulder

27. Terms of relative position are used to describe
   a. the relationship of siblings within a family.
   b. the importance of the various functions of organ systems in maintaining life.
   c. the location of one body part with respect to another.

28. A sagittal section divides the body into
   a. superior and inferior portions.
   b. right and left portions.
   c. anterior and posterior portions.

29. The terms *epigastric*, *hypochondriac*, and *iliac* are examples of _____ _____ .

Study Activities

I. Aids to Understanding Words
Define the following word parts. (p. 7)

append-

cardi-

cran-

dors-

homeo-

-logy

meta-

pariet-

pelv-

peri-

pleur-

-stasis

-tomy

II. Anatomy and Physiology (p. 9)

Explain how the structure of the fingers is related to their grasping function. (p. 9)

III. Characteristics of Life (p. 9)

Describe the following characteristics of life. (p. 9)

movement

responsiveness

growth

reproduction

respiration

digestion

absorption

circulation

assimilation

excretion

IV. The Maintenance of Life (pp. 10–11)

A. Match the terms in the first column with the second-column statements that define their role in the maintenance of life. (p. 10)

<u>a</u> 1. water   a. essential for metabolic processes

<u>e</u> 2. food   b. governs rate of chemical reactions

<u>d</u> 3. oxygen   c. creates a pressing or compressing action

<u>b</u> 4. heat   d. necessary for release of energy

<u>c</u> 5. pressure   e. provides chemicals for building new living matter

B. Describe the homeostatic mechanisms that control temperature; blood pressure. (pp. 10–11)

V. Levels of Organization (p. 11)

Arrange the following structures in increasing levels of complexity: organ systems, organelles, organism, organs, cells, tissues. (p. 11)

VI. Body Cavities (pp. 12–13)

A. The dorsal cavity is subdivided into the _____ cavity and the _____ cavity. (p. 12)

B. Answer the following concerning the ventral cavity. (p. 12)
1. The ventral cavity is subdivided into the _____ cavity and the _____ cavity.
2. The _____ divides the ventral cavity.
3. List the viscera found in each of the portions of the ventral cavity.

C. The pelvic cavity is enclosed by the _____ _____ . (p. 12)

VII. Thoracic and Abdominopelvic Membranes (p. 14)

A. Fill in the blanks.

1. The walls of the thoracic cavity are lined with a membrane called the _____

_____ .

2. The lungs are covered by the _____ _____ .

3. Why is the pleural cavity called a potential space?

B. Name and describe the membranes covering the heart.

C. The linings of the abdominopelvic cavity are the _____ _____ and the

_____ _____ .

VIII. Organ Systems

Fill in the following chart. (pp. 14–16)

Structure and function of organ systems

Function	Organ System	Organs in System
Support and Movement	1.	1.
	2.	2.
Integration and Coordination	1.	1.
	2.	2.
Processing and Transporting	1.	1.
	2.	2.
	3.	3.
	4.	4.
	5.	5.
Reproduction: Female	1.	1.
Male	2.	2.

IX. Anatomical Terminology (pp. 16–20)

A. Using this illustration, specify the terms that describe the relationship of one point on the body to another. (pp. 16–17)

1. Point (*a*) in relation to point (*d*).

    Superior

2. Point (*f*) in relation to point (*h*).

    proximal

3. Point (*g*) in relation to point (*i*).

    medial

4. Point (*l*) in relation to point (*j*).

    distal

5. Point (*i*) in relation to point (*g*).

    lateral

6. Point (*c*) in relation to point (*a*).

    inferior

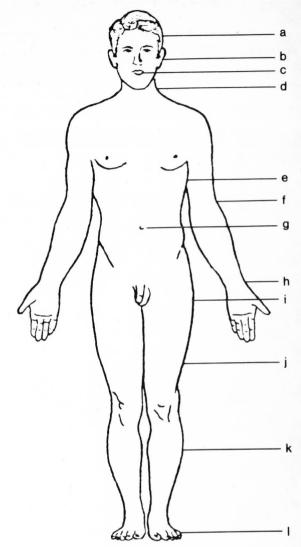

B. Using this illustration, perform the following exercises. (pp. 17–20)

1. Draw a line through the drawing to indicate a midsagittal section. How is this different from a frontal section?

2. Draw a line through the drawing to indicate a transverse section.
3. Define *cross section, longitudinal section,* and *oblique section.*

4. Locate and label the following body regions: epigastric, umbilical, hypogastric, hypochondriac, lumbar, and iliac. Locate these regions on yourself or on a partner.

5. Locate and label the following body parts on the diagram: antebrachium, antecubital, axillary, brachial, buccal, cervical, groin, inguinal, mammary, ophthalmic, palmar, pectoral.

When you have finished the study activities to your satisfaction, retake the mastery test and compare your results with your initial attempt. If you are not satisfied with your performance, repeat the appropriate study activities.

2 Chemical Basis of Life

Overview

This chapter introduces some basic concepts of chemistry, a science that studies the composition of substances and the changes that occur as basic elements combine. It explains how substances combine to make up matter (objectives 1–5), how substances are classified as acid or base (objective 6), and the organic and inorganic substances that make up the living cell (objectives 7 and 8).

Knowledge of basic chemical concepts enhances understanding of the functions of cells and of the human body.

Chapter Objectives

After you have studied this chapter, you should be able to

1. Explain how the study of living material is dependent on the study of chemistry.
2. Describe the relationships between matter, atoms, and molecules.
3. Discuss how atomic structure is related to ways in which atoms interact.
4. Explain how molecular and structural formulas are used to symbolize composition of compounds.
5. Describe two types of chemical reactions.
6. Discuss the concept of pH.
7. List the major groups of inorganic substances that are common in cells.
8. Describe the general roles played in cells by various types of organic substances.

Focus Question

How is chemistry related to the structure and function of living things and their parts?

Mastery Test

Now take the mastery test. Do not guess. As soon as you complete the test, correct it. Note your successes and failures so that you can read the chapter to meet your learning needs.

1. The branch of science that deals with the composition of substances is _____ .

2. What is matter? In what forms can it be found?

3. The substances that compose all matter are called _____ .

4. What elements are most plentiful in the human body?

5. An atom is made up of
   a. a nucleus.
   b. protons.
   c. neutrons.
   d. electrons.
   e. all of the above.

6. Match the following.
   ____ a. neutron
   ____ b. proton
   ____ c. electron

   1. positive electrical charge
   2. negative electrical charge
   3. no electrical charge

7. The atomic number of an element is determined by the number of _____ .

8. When atoms combine, they gain or lose
   a. electrons.
   b. neutrons.
   c. protons.
   d. nuclei.

9. An element is inactive if _____ .

10. An ion is
    a. an atom that is electrically charged.
    b. an atom that has gained an electron.
    c. an atom that has lost an electron.
    d. all of the above.

11. An electrovalent bond is created by
    a. a positive and a negative ion attracting each other.
    b. two or more positive ions combining.
    c. two or more negative ions combining.

12. In forming a covalent bond, electrons are
    a. shared by two atoms.   c. taken up by an atom.
    b. given up by an atom.   d. none of the above.

13. A compound is formed when atoms of _____ elements combine.

14. $C_6H_{12}O_6$ is an example of a(n) _____ formula.

15. 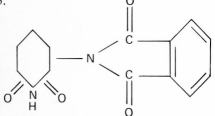 is an example of a(n) _____ formula.

16. Two major types of chemical reactions are called _____ and _____ .

17. The symbol $\rightleftharpoons$ indicates a _____ reaction.

18. A(n) _____ accepts hydroxide ions in water.

19. A(n) _____ releases hydrogen ions in water.

20. What is the pH of a neutral solution?

21. Identify the following cell constituents with an O if they are organic and an I if they are inorganic.
    a. water (   )   d. oxygen (   )
    b. carbohydrate (   )   e. protein (   )
    c. glucose (   )   f. fats (   )

22. Triglycerides, phospholipids, and steroids are important _____ found in the human cell.

23. The function of nucleic acids is
    a. to store information and control life processes.
    b. to act as receptors for hydrogen ions.
    c. to neutralize bases within the cell.

Study Activities

I. Aids to Understanding Words

Define the following word parts. (p. 35)

di-	mono-
glyc-	poly-
lip-	sacchar-
-lyt	syn-

II. Structure of Matter (pp. 36–42)

A. Answer these questions concerning elements and atoms. (p. 36)

1. Anything that has weight and takes up space is _____ .

2. Basic substances are called _____ .

3. Tiny, invisible particles that make up basic substances are called _____ .

4. Two or more particles of the same basic substance form a(n) _____ .

5. Two or more particles of different basic substances form a(n) _____ .

B. Fill in the chart. (p. 36)

Element	Symbol	Element	Symbol
Oxygen		Sodium	
Carbon		Magnesium	
	H	Cobalt	
Nitrogen			Cu
	Ca		F
	P	Iodine	
	K		Fe
Sulfur			Mn
	Cl	Zinc	

C. Answer the questions that pertain to the accompanying diagram. (pp. 36–37)

1. What is the atomic number of this atom?

2. How many electrons are needed to fill its outer shell?

3. Is this element active or inert?

4. Identify this element.

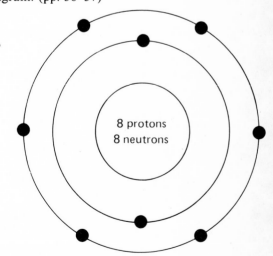

8 protons
8 neutrons

D. Draw diagrams of an electrovalent and a covalent bond. (pp. 38–39)

How are these bonds different?

E. Label the following chemical reactions. (p. 41)

   1. A + B → AB _____

   2. AB → A + B _____

   3. A + B ⇌ AB _____

F. Answer the questions that pertain to the accompanying diagram. (p. 40)

```
        H    O
         \  //
          C
          |
     H — C — O — H
          |
 H — O — C — H
          |
     H — C — O — H
          |
     H — C — O — H
          |
     H — C — O — H
          |
          H
```

   1. Give the molecular formula of this substance.

   2. Identify this compound.

   3. Is it an organic or an inorganic substance? Why?

G. Answer these questions concerning acid and base concentrations. (pp. 41–42)
   1. What is meant by pH?

   2. Substances that release hydrogen ions (H^+) in water are called _____ .

   3. Substances that release ions that combine with hydrogen ions are _____ .

   4. Identify these substances as either acid or base.
   carrot: pH 5.0   tomato: pH 4.2
   milk of magnesia: pH 10.5   lemon: pH 2.3
   human blood: pH 7.4   distilled water: pH 7.0

III. Chemical Constituents of Cells (pp. 42–46)

A. What roles do the following inorganic substances play in the cell? Be specific. (p. 42)
   water   carbon dioxide

   oxygen   inorganic salts (Na^+, K^+, Ca^{+2}, HCO_3^-, PO_4^{-3})

B. Answer these questions concerning carbohydrates. (p. 43)
   1. What is the role of carbohydrates in maintaining the cell?

   2. The form of carbohydrate utilized by the cell is _____ .

   3. The form in which humans store carbohydrates is _____ .
C. Answer these questions concerning lipids. (pp. 44–45)
   1. What is the role of lipids in maintaining the cell?

   2. Fats are composed of _____ , _____ ,
   and _____ .

   3. Fats containing single carbon-carbon bonds are _____ .

   4. Fats containing one or more double-bonded carbon atoms are _____ .
   5. Describe the molecular structure of these lipids.
   triglycerides

   phospholipids

   steroids

   6. Identify the characteristics and functions of these lipids.
   triglycerides

   phospholipids

   steroids

D. Answer these questions concerning proteins. (pp. 44–46)

    1. What is the role of protein in maintaining the cell?

    2. The building materials of a protein are _____ _____ .

    3. A protein molecule that has become disorganized and lost its shape is said to be _____ .

E. Nucleic Acids (p. 46)

    1. The nucleic acids are _____ and _____ .

    2. What is their role in cell function?

When you have finished the study activities to your satisfaction, retake the mastery test and compare your results with your initial attempt. If you are not satisfied with your performance, repeat the appropriate study activities.

3 Cells

Overview

This chapter deals with the structural unit of the human body—the cell. It explains how cells vary from each other, the makeup of a composite cell, and the contribution of each of the organelles to cellular function (objectives 1–4). It introduces the various mechanisms used to transport material into and out of the cell (objective 6). The cell nucleus and its parts are described (objective 5). The life cycle of a cell, including the processes of cell reproduction, is explained (objectives 7 and 8).

An understanding of cell structure is basic to understanding how cells support life at the cellular and organismic levels.

Chapter Objectives

After you have studied this chapter, you should be able to

1. Explain how cells vary from one another.
2. Describe the general characteristics of a composite cell.
3. Explain how the structure of a cell membrane is related to its function.
4. Describe each kind of cytoplasmic organelle and explain its function.
5. Describe the cell nucleus and its parts.
6. Explain how substances move through cell membranes.
7. Describe the life cycle of a cell.
8. Explain how a cell reproduces.

Focus Question

How does the structure of cellular organelles contribute to and support the functions of the organelle and, in turn, the cell?

Mastery Test

Now take the mastery test. Do not guess. As soon as you complete the test, correct it. Note your successes and failures so that you can read the chapter to meet your learning needs.

1. The cells of the human body vary in terms of size and _____ .

2. Which of the following statements about a hypothetical composite cell are true?
   a. It is necessary to construct a composite cell because cells vary so much, based on their function.
   b. It contains structures that occur in many kinds of cells.
   c. It contains only structures that occur in all cells although the characteristics of the structure may vary.
   d. It is an actual cell type chosen because it occurs most commonly in the body.

3. The two major portions of the cell, each surrounded by a membrane, are the _____ and the _____ .

4. The organelles are located in
   a. the nucleolus.
   b. the cytoplasm.
   c. the cell matrix.
   d. the cell membrane.

5. The cell membrane allows some substances to pass through it and excludes others. This is possible because the cell membrane is _____ _____ .

6. The cell membrane is composed of a double layer of
   a. protein molecules.
   b. phospholipid molecules.
   c. polysaccharide molecules.
   d. amino acids.

7. The organelle that functions as a system of transport for materials from one part of the cytoplasm to another is the _____ _____ .

8. Ribosomes function in the synthesis of protein molecules. (true, false)

9. The Golgi apparatus is involved in the "packaging" of proteins for secretion to the (inside, outside) of the cell.

10. The mitochondria function in the release of _____ to the cells.

11. The enzymes of the lysosome function to
    a. control cell reproduction.
    b. digest bacteria and damaged cell parts.
    c. release energy from its storage place within the cell.
    d. control the Krebs cycle.

12. Which of the following statements about the centrosome are true?
    a. It is located near the nucleus.
    b. The centrioles of the centrosome function solely in reproduction.
    c. The centrosome is concerned with the distribution of chromosomes.
    d. All of the above are true.

13. Cilia are found on the surface of
    a. endothelial cells.
    b. epithelial cells.
    c. epidermal cells.
    d. mucosa.

14. Microfilaments are rods of protein involved in cellular _____ .

15. The structures that float in the nucleoplasm of the nucleus are the _____ and the _____ .

16. The process that allows moving of gases and ions from areas of higher concentration to areas of lower concentration until equilibrium has been achieved is called _____ .

17. The process by which nonsoluble material moves through the cell membrane by using a carrier molecule is called _____ _____ .

18. The process by which water moves across a semipermeable membrane from areas of low concentration of solute to areas of higher concentration is called _____ .

19. A hypertonic solution is one that
    a. contains a greater concentration of solute than the cell.
    b. contains the same concentration of solute as the cell.
    c. contains a lesser concentration of solute than the cell.

20. The process by which molecules are forced through a membrane by pressure that is greater on one side than on the other side is called _____ .

21. The process that uses energy to move ions across a concentration gradient from an area of lower concentration to an area of higher concentration is called _____ _____ .

22. The process by which cells engulf water molecules is called _____ .

23. A process that allows cells to take in molecules of solids is called _____ .

24. Once solid material is taken into a vacuole, which of the following statements best describes what happens?
    a. A ribosome enters the vacuole and uses the amino acids in the invader to form new protein.
    b. A lysosome combines with the vacuole and digests the enclosed solid material.
    c. The vacuole remains separated from the cytoplasm and the solid material persists unchanged.
    d. Oxygen enters the vacuole and burns the enclosed solid material.

25. The process that ensures duplication of DNA molecules during cell reproduction is _____ .

26. Match these events with the correct description.
    ____ a. prophase
    ____ b. metaphase
    ____ c. anaphase
    ____ d. telophase

    1. Microtubules shorten; chromosomes pulled toward centrioles.
    2. Chromatin forms chromosomes; nuclear membrane and nucleolus disappear.
    3. Chromosomes elongate; nuclear membranes form around each chromosome set.
    4. Chromosomes become arranged midway between centrioles; duplicate parts of chromosomes become separated.

27. The process by which cells develop unique characteristics in structure and function is called

    _____ .

Study Activities

I. Aids to Understanding Words

Define the following word parts. (p. 51)

cyt-   iso-
endo-   mit-
hyper-   phag-
hypo-   pino-
inter-   -som

II. A Composite Cell (pp. 52–58)

Fill in the following chart. (pp. 54–58)

Structure and function of cellular organelles

Organelle	Structure	Function
Cell membrane		
Endoplasmic reticulum		
Ribosomes		
Golgi apparatus		
Mitochondria		
Lysosomes		
Centrosome		
Vesicles		
Cilia and flagella		
Microfilaments and microtubules		
Nuclear envelope		
Nucleolus		
Chromatin		

III. Movements through Cell Membranes (pp. 58–63)

A. What substances cross the cell membrane? (pp. 58–63)

B. Answer these questions concerning movement through cell membranes. (pp. 58–61)
   1. In what direction do the molecules of solute move in diffusion?

   2. When does diffusion stop?

   3. What substances in the human body are transported by diffusion?

   4. Describe facilitated diffusion.

   5. How does osmosis differ from diffusion?

C. Below are three drawings of a red blood cell in solutions of varying tonicity. Label the tonicity of the solution in each drawing and explain what is happening and why. (p. 61)

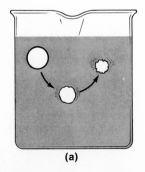
(a)

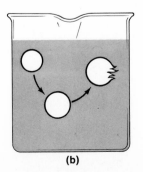
(b)

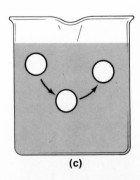
(c)

D. Answer these questions concerning the accompanying illustration. (p. 61)
    1. What process is illustrated here?

    2. What provides the force needed to pull the liquid through the solids?

    3. Where within the body does this process occur?

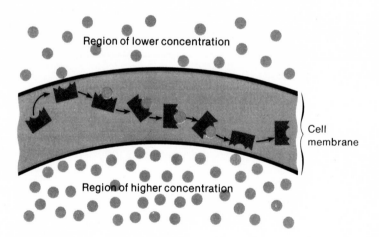

E. Answer the questions concerning the accompanying diagram. (pp. 62–63)

    1. What transport mechanism is illustrated here?

    2. What provides the necessary force for this process?

    3. What is the source of this force?

    4. How are molecules transported across the cell membrane?

    5. What substances are transported by this mechanism?

F. Answer the questions concerning the accompanying diagram. (p. 63)

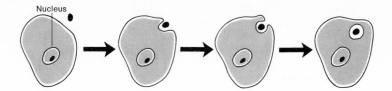

Nucleus

1. What mechanism is illustrated in this drawing?

2. What organelles are involved?

3. What kinds of foreign objects are transported?

4. How is this mechanism important to cell survival?

IV. Life Cycle of a Cell (pp. 63–68)

A. The series of changes that a cell undergoes from its formation until its reproduction is called its
_____ . (p. 63)

B. Describe each of the following events in mitosis. (pp. 64–66)
prophase

metaphase

anaphase

telophase

C. Describe the events of cytoplasmic division. (p. 66)

D. Answer these questions concerning cell differentiation. (pp. 66–68)
  1. The process by which cells develop differences in structure and function is
  _____ .

  2. Describe the role of DNA in this process.

When you have completed the study activities to your satisfaction, retake the mastery test and compare your
performance with your initial attempt. If there are still areas you do not understand, repeat the appropriate study
activities.

4 Cellular Metabolism

Overview

This chapter deals with two basic cellular processes: the use of energy and the use of genetic information to control cell processes. Specifically, this chapter discusses how enzymes control cell metabolism, how energy is released and made available to the cell, and how carbohydrates, lipids, and proteins are metabolized (objectives 1–5). It also explains how genetic information is stored, how it is used to control cell processes, and how the cell changes when genetic information is altered (objectives 6, 7, and 8).

An understanding of life processes at the cellular level is basic to understanding how life processes occur at more complex levels—as in the whole organism.

Chapter Objectives

After you have studied this chapter, you should be able to

1. Define *anabolic* and *catabolic metabolism.*
2. Explain how enzymes control metabolic processes.
3. Explain how chemical energy is released by respiratory processes.
4. Describe how energy is made available for cellular activities.
5. Describe the general metabolic pathways of carbohydrates, lipids, and proteins.
6. Explain how genetic information is stored within nucleic acid molecules.
7. Explain how genetic information is used in the control of cellular processes.
8. Describe how DNA molecules are replicated.

Focus Question

How do cells carry on life processes?

Mastery Test

Now take the mastery test. Do not guess. As soon as you complete the test, correct it. Note your successes and failures so that you can read the chapter to meet your learning needs.

1. The metabolic process that synthesizes materials needed for cellular growth is called _____

   _____ .

2. The metabolic process that breaks down complex molecules into simpler ones is called _____

   _____ .

3. The process by which two molecules are joined together to form a more complex molecule is called
   a. dehydration synthesis.   c. atomization.
   b. chemical bonding.

4. Glycerol and fatty acids become bonded to form water and
   a. cholesterol.   c. fat molecules.
   b. lard.   d. wax.

5. Amino acids are joined by a peptide bond to form water and _____ .

6. The process by which water is added to a complex molecule to break it down, as represented by the equation
   $C_{12}H_{22}O_{11} + H_2O = C_6H_{12}O_6 + C_6H_{12}O_6$, is called _____ or _____ .

7. Enzymes are composed of
   a. lipid.   c. protein.
   b. carbohydrate.   d. inorganic salts.

8. A substance that initiates or speeds up chemical reactions without being altered chemically is a(n)

   _____ .

9. An enzyme acts only on a particular substance that is called a
   a. binding site.   c. complement.
   b. substrate.   d. histologue.

10. An enzyme's ability to recognize the substance upon which it will act seems to be based on
    a. atomic weight.   c. structural formula.
    b. molecular shape.

11. A substance needed to convert an inactive form of an enzyme to an active form is called a(n)

    _____ .

12. The form of energy utilized by most cellular processes is
    a. chemical.   c. thermal.
    b. electrical.   d. mechanical.

13. The process by which energy is released in the cell is called _____ .

14. The initial phase of respiration that occurs in the cytoplasm and produces two 3-carbon pyruvic acid molecules
    plus energy is called _____ _____ .

15. The second phase of respiration that occurs in the mitochondria and produces carbon dioxide, water, and energy
    is called _____ _____ .

16. What element is needed for this second phase to take place? _____

17. Name the storage place for the energy released by cellular respiration. _____

18. A particular sequence of enzyme-controlled reactions is called a(n) _____ _____ .

19. Carbohydrate must be converted to a(n) _____ to be available as an energy source.

20. Fats contain (more, less) energy per gram compared to carbohydrates.

21. If proteins are to be used as an energy source, they undergo a process in the liver called

    _____ .

22. What element is removed in this process?
    a. carbon   c. oxygen
    b. hydrogen   d. nitrogen

23. We inherit traits from our parents because
    a. DNA contains genes that are the carriers of   c. our species, *Homo sapiens,* reproduces sexually.
       inheritance.
    b. genes tell the cells to construct protein in a
       unique way for each individual.

24. A molecule consisting of a double spiral with sugar and phosphates forming the outer strands and organic bases
    joining the two strands is _____ .

25. DNA molecules are located in the _____ . Protein synthesis takes place in the

    _____ .

26. Two types of RNA are _____ RNA and _____ RNA.

27. The function of RNA is
    a. the formation of lipids, such as cholesterol.   c. to control the bonding of amino acids.
    b. to guide the breakdown of polysaccharides.

28. When the genetic material of a cell is altered, the result may be a(n) _____ .

Study Activities

I. Aids to Understanding Words

Define the following word parts. (p. 73)

an-   de-

ana-   mut-

cata-   zym-

II. Metabolic Processes (p. 74)

A.   Answer these questions concerning anabolic metabolism. (p. 74)
1.   Define anabolic metabolism.

2.   The following formula is an example of anabolic metabolism. Label the formula and identify the process illustrated.

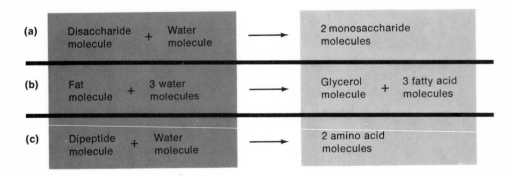

B.   Answer these questions concerning catabolism. (p. 74)
1.   Define *catabolism*.

2.   What process is illustrated here?

(a)	Disaccharide molecule + Water molecule	→	2 monosaccharide molecules
(b)	Fat molecule + 3 water molecules	→	Glycerol molecule + 3 fatty acid molecules
(c)	Dipeptide molecule + Water molecule	→	2 amino acid molecules

III. Control of Metabolic Reactions (pp. 76–77)

A.   Enzymes promote chemical reactions in cells by _____ the amount of _____ needed to initiate a reaction. (p. 76)

B.   What is the relationship between an enzyme and a substrate? Explain how this relationship works. (p. 76)

C. What is a coenzyme? What substances are coenzymes? (p. 76)

IV. Energy for Metabolic Reactions (pp. 77–78)

    A. Answer these questions concerning energy and its release. (p. 77)
        1. What is energy?

        2. List six common forms of energy.

        3. The form of energy used by cell processes is _____ .

        4. This energy is released by the process of _____ .

    B. Fill in the following chart, comparing aerobic and anaerobic respiration. (pp. 77–78)

Types of respiration

Type	Anaerobic	Aerobic
Amount of energy used		
End product of oxidation		
Location of reaction within cell		
How released energy is captured		
Number of molecules formed		

V. Metabolic Pathways (pp. 78–81)

    A. What is a metabolic pathway? (p. 78)

    B. The metabolic pathways for carbohydrates, lipids, and proteins are such that an excess of any of these substances can be stored as _____ . (pp. 78–81)

VI. Nucleic Acids and Protein Synthesis (pp. 82–86)

    A. Answer these questions concerning genes. (p. 82)
        1. What is a gene?

        2. How are genes necessary to cell metabolism?

B. What are the possible bases of nucleotides in DNA? (p. 82)

C. What is the structure of DNA? (p. 83)

D. Describe the roles of messenger RNA and transfer RNA in protein synthesis. (p. 86)

E. Draw the matching strand of DNA for the strand illustrated here. (p. 86)

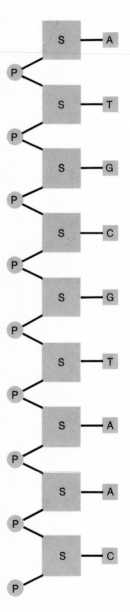

F. Describe the replication of DNA. (p. 86)

When you have completed the study activities to your satisfaction, retake the mastery test and compare your performance with your initial attempt. If there are still areas you do not understand, repeat the appropriate study activities.

5 Tissues

Overview

This chapter deals with the simplest level of organization of cells—tissues. It explains the types of tissue that occur in the human body, the general functions of each of these types of tissue, and the organs in which the various types of tissue occur (objectives 1–9).

   The characteristics of a tissue remain the same regardless of where it occurs in the body. Knowledge of these characteristics is basic to understanding how a specific tissue contributes to the function of an organ.

Chapter Objectives

After you have studied this chapter, you should be able to

1.   Describe the general characteristics and functions of epithelial tissue.
2.   Name the types of epithelium and identify an organ in which each is found.
3.   Explain how glands can be classified.
4.   Describe the general characteristics of connective tissue.
5.   Describe the major cell types and fibers of connective tissue.
6.   List the major types of connective tissues that occur within the body.
7.   Describe the major functions of each type of connective tissue.
8.   Distinguish between the three types of muscle tissue.
9.   Describe the general characteristics and functions of nerve tissue.

Focus Question

How is tissue related to the organization of the body?

Mastery Test

Now, take the mastery test. Do not guess. As soon as you complete the test, correct it. Note your successes and failures so that you can read the chapter to meet your learning needs.

1.   Cells in a tissue are (similar, dissimilar).

2.   The function of epithelial tissue is to
   a. support body parts.   c. bind body parts together.
   b. cover body surfaces.   d. form the framework of organs.

3.   Which of the following statements about epithelial tissue is(are) true?
   a. Epithelial tissue has no blood vessels.
   b. Epithelial cells reproduce slowly.
   c. Epithelial cells are nourished by substances diffusing from connective tissue.
   d. Injuries to epithelial tissue bleed profusely.

4.   Match the following types of epithelial cells with the correct location.
   ____ a. simple squamous epithelium   1. lining of the ducts of salivary glands
   ____ b. simple cuboidal epithelium   2. lining of respiratory passages
   ____ c. simple columnar epithelium   3. epidermis of the skin
   ____ d. pseudostratified columnar epithelium 4. air sacs of lungs, walls of capillaries
   ____ e. stratified squamous epithelium   5. lining of digestive tract

5.   The inner lining of the urinary bladder and the passageways of the urinary tract are composed of

   _____   _____ .

6. A gland which secretes its products into ducts opening into an external or internal surface is called a(n) _____ gland.

7. The functions of connective tissue are
   a. support.
   b. protection.
   c. covering.
   d. fat storage.

8. The major structural protein of the body and white connective tissue is _____.

9. Yellow connective tissue that can be stretched and returned to its original shape is _____.

10. Which of the following statements is(are) true of loose connective tissue?
    a. It forms heavy, tough membranes under the skin and between muscles.
    b. It contains both yellow and white fibers.
    c. It has a meager blood supply.
    d. It contains few fibroblasts.

11. Which of the following statements is(are) true about adipose tissue?
    a. It is a specialized form of loose connective tissue.
    b. It occurs around the kidneys, behind the eyeballs, and around various joints.
    c. It serves as a conserver of body heat.
    d. It serves as a storehouse of energy for the body.

12. Tendons connect _____ to _____ .

    Ligaments connect _____ to _____ .

    Both are examples of _____ connective tissue.

13. The most rigid connective tissue is _____ .

14. The intercellular material of vascular tissue is _____ .

15. The cells of reticuloendothelial tissue are usually
    a. phagocytic.
    b. rigid.
    c. proliferative.
    d. almost inert.

16. The three types of muscle tissue are _____ , _____ , and _____ .

17. Coordination and regulation of body functions is the function of _____ tissue.

Study Activities

I. Aids to Understanding Words

Define the following word and word parts. (p. 91)

adip-   macro-

-cyt   oss-

epi-   pseudo-

-glia   squam-

inter-   stratum

II. Epithelial Tissues (pp. 92–96)

A. List five functions of epithelial tissue. (p. 92)

B. Answer these questions concerning simple squamous epithelium. (p. 92)
   1. The structure of simple squamous epithelium is _____ .

2. Simple squamous epithelium is found where _____ and _____ take place.

C. Answer these questions concerning simple cuboidal epithelium. (p. 93)
   1. Describe the structure of simple cuboidal epithelium.

   2. Where is this type of tissue found?

   3. The function of simple cuboidal epithelium is _____ and _____ .

D. Answer these questions concerning pseudostratified columnar epithelium. (p. 94)
   1. Microscopic, hairlike projections called _____ are a characteristic of columnar epithelium.
   2. Where is this tissue found?

E. Answer these questions concerning stratified squamous epithelium. (p. 94)
   1. Describe the structure of stratified squamous epithelium.

   2. Where is this tissue found?

F. What is the special characteristic of transitional epithelium? (p. 95)

G. Describe glandular epithelium. (p. 96)

H. Fill in the following chart. (p. 96)

Types of glandular secretions

Type of Gland	Description of Secretion	Examples
Merocrine		
Apocrine		
Holocrine		

III. Connective Tissues (pp. 97–103)

A. What are the functions of connective tissue? (p. 97)

B. How do yellow and white connective tissue differ? (p. 97)

C. Where is adipose connective tissue found, and what is its function? (p. 98)

D. What is the difference between a ligament and a tendon? (p. 99)

E. Fill in the following chart. (pp. 99–100)

Types of cartilage

Type	Location	Function
Hyaline		
Elastic		
Fibrocartilage		

F. Answer these questions concerning bone. (pp. 100–102)
    1. What are the characteristics of bone?

    2. Bone injuries heal relatively rapidly. Why is this true?

G. Answer these questions concerning blood (vascular connective tissue). (p. 102)
    1. What is the intercellular material of vascular connective tissue?

    2. What cells are found in the intercellular material?

H. Answer these questions concerning reticuloendothelial tissue. (p. 103)
    1. Where is reticuloendothelial tissue found?

    2. What is the most common type of reticuloendothelial cell? What is its function?

IV. Muscle Tissues (pp. 103–4)

A. What are the characteristics of muscle tissue?

B. Fill in the following chart.

Muscle tissue

Type	Structure	Control	Location
Skeletal			
Smooth			
Cardiac			

V. Nerve Tissue (pp. 104–5)

    A.   What is the basic cell of nerve tissue?

    B.   What is the function of neuroglial cells in nerve tissue?

    C.   What is the function of nerve tissue?

When you have completed the study activities to your satisfaction, retake the mastery test and compare your performance with your initial attempt. If there are still areas you do not understand, repeat the appropriate study activities.

6 Skin and Integumentary System

Overview

This chapter describes the skin and its accessory organs. It explains the structure and function of the layers of the skin (objectives 1–3), skin color (objective 4), and the structure of hair, nails, and sweat glands (objective 5). It tells how the skin helps to regulate body temperature (objective 6).

Study of the integumentary system is essential to understanding how the body controls interaction between the internal and external environments.

Chapter Objectives

After you have studied this chapter, you should be able to

1. Describe the four major types of membranes.
2. Describe the structure of the various layers of the skin.
3. List the general functions of each of these layers of skin.
4. Summarize the factors that determine skin color.
5. Describe the accessory organs associated with the skin.
6. Explain how the skin functions in regulating body temperature.

Focus Question

How do the structure and function of the skin earn it the title of *the body's first line of defense?*

Mastery Test

Now take the mastery test. Do not guess. As soon as you complete the test, correct it. Note your successes and failures so that you can read the chapter to meet your learning needs.

1. List the four major types of membrane.

2. Mucous membranes are located
   a. around the organs of the respiratory system.
   b. in areas where two surfaces may rub together.
   c. in the lining of cavities and tubes that have openings to the outside of the body.

3. Serous membranes are located
   a. around the structures of the dorsal cavity.
   b. in body cavities that are completely closed to the outside of the body.
   c. wherever two bones come together.

4. The function of synovial membranes is to secrete a fluid that
   a. nourishes surrounding tissue.
   b. provides insulation.
   c. reduces friction.
   d. none of the above.

5. The outer layer of skin is called the _____ .

6. The inner layer of skin is called the _____ .

7. The masses of connective tissue beneath the inner layers of skin are called the _____

   _____ .

8. The cells of the skin that reproduce are in the
   a. keratin.
   b. stratum corneum.
   c. stratum germinativum.
   d. epidermis.

9. The pigment that helps protect the deeper layers of the epidermis is
   a. melanin.
   b. trichosiderin.
   c. biliverdin.
   d. bilirubin.

10. Blood vessels supplying the skin are located in the _____ .

11. The subcutaneous layer functions as a(n) _____ _____ .

12. The glands usually associated with hair follicles are
   a. apocrine glands.
   b. endocrine glands.
   c. sebaceous glands.
   d. exocrine glands.

13. Where are the sweat glands most numerous?
   a. the face
   b. the groin and the axilla
   c. the palms and soles
   d. evenly distributed over the body surface

14. The sweat glands associated with regulation of body temperature are the
   a. endocrine glands.
   b. eccrine glands.
   c. exocrine glands.
   d. apocrine glands.

15. Which of the following organs produces the most heat?
   a. kidneys
   b. bones
   c. muscles
   d. lungs

Study Activities

I. Aids to Understanding Words

Define the following word parts. (p. 111)

cut-

derm-

epi-

follic-

kerat-

melan-

seb-

II. Types of Membranes (p. 112)

List and locate the four major types of membranes.

III. Skin and Its Tissues (pp. 112–15)

A.   List the functions of the skin. (p. 112)

B.   What kinds of tissue are found in the skin? (p. 112)

C. Answer the following questions about the layers of the epidermis. (pp. 112–13)
   1. What is the function of the stratum germinativum?

   2. What is the function of the stratum corneum?

   3. A callus or a corn is the result of a(n) _____ in cell reproduction in response to
      _____ or _____ .
D. Deep layers of skin are protected from the ultraviolet portion of sunlight by _____ .
   (p. 113)

E. What environmental factors influence skin color? (p. 114)

F. Describe cutaneous carcinomas and cutaneous melanomas. (p. 114)

G. Describe the structure and function of the dermis and the subcutaneous layer. (p. 115)

IV. **Accessory Organs of the Skin** (pp. 115–17)

   A. How is hair formed in this follicle? (p. 115)

   B. Describe how hair responds to cold temperature or strong emotion. (p. 115)

   C. Where are sebaceous glands located, and what is the function of the substance they secrete? (p. 116)

D. Label the following parts of a hair follicle on the illustration below: hair shaft, hair follicle, region of cell division, arrector pili muscle, sebaceous gland. (p. 116)

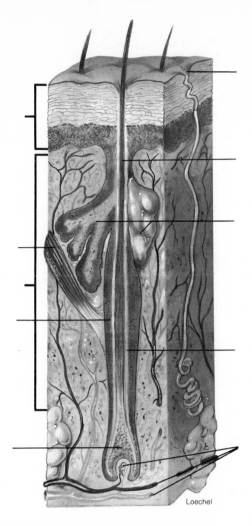

Loechel

E. Where is the growing portion of the nail located? (p. 116)

F. Compare apocrine and eccrine sweat glands in relation to location, association with other skin structures, and activating stimuli. (p. 117)

V. Regulation of Body Temperature (pp. 118–19)

Describe the roles of the nervous, muscular, circulatory, and respiratory systems in heat regulation.

When you have completed the study activities to your satisfaction, retake the mastery test and compare your performance with your initial attempt. If there are still areas you do not understand, repeat the appropriate study activities.

7 Skeletal System

Overview

This chapter deals with the skeletal system—the bones that form the framework for the body. It explains the function and structure of bones (objectives 1 and 3). The development of different types of bone is also explained (objective 2). The chapter describes skeletal organization and the location of specific bones within various parts of the skeleton (objectives 4 and 5). Various types of joints and the movements made possible by these joints (objectives 6, 7, and 8) are also described in chapter 7.

Movement is a characteristic of living things. A study of the skeletal system is necessary to understand how complex organisms, such as humans, are organized to accomplish movement.

Chapter Objectives

After you have studied this chapter, you should be able to

1. Describe the general structure of a bone and list the functions of its parts.
2. Distinguish between intramembranous and endochondral bones and explain how such bones develop and grow.
3. Discuss the major functions of bones.
4. Distinguish between the axial and appendicular skeletons and name the major parts of each.
5. Locate and identify the bones and the major features of the bones that comprise the skull, vertebral column, thoracic cage, pectoral girdle, upper limb, pelvic girdle, and lower limb.
6. List three types of joints, describe their characteristics, and name an example of each.
7. List six types of freely movable joints and describe the actions of each.
8. Explain how skeletal muscles produce movements at joints and identify several types of such movements.

Focus Question

Is bone an essentially inanimate framework upon which muscle works to achieve movement, or does it have an active role in maintaining the internal environment?

Mastery Test

Now take the mastery test. Do not guess. As soon as you complete the test, correct it. Note your successes and failures so that you can read the chapter to meet your learning needs.

1. The shaft of a long bone is the
   a. epiphysis.
   b. diaphysis.

2. Bone that consists of tightly packed tissue is called _____ .

3. Bone that consists of numerous bony bars and plates separated by irregular spaces is called _____ .

4. The medullary cavity of a long bone is filled with _____ .

5. Bones that develop from layers of membranous connective tissue are called _____

   _____ .

6. Bones that develop from masses of hyaline cartilage are called _____

   _____ .

7. The band of cartilage between the primary and secondary ossification centers in long bones is called the
   a. osteoblastic band.   c. periosteal plate.
   b. calcium disk.   d. epiphyseal disk.

8. To accomplish movement, bones and muscles function together to act as _____ .

9. Which of the following bones contain red marrow for blood cell formation in a healthy adult?
   a. pelvis
   b. small bones of the wrist
   c. ribs
   d. shaft of long bones

10. Which of the following inorganic salts are stored in bone?
    a. potassium
    b. calcium
    c. lead
    d. chlorine

11. The usual number of bones in the human skeleton is _____ .

12. List the major parts of the axial skeleton.

13. List the major parts of the appendicular skeleton.

14. The only movable bone of the skull is the
    a. nasal bone.
    b. mandible.
    c. maxilla.
    d. vomer.

15. The bone that forms the back of the skull and joins the skull along the lamboidal suture is the
    _____ bone.

16. The upper jaw is formed by the _____ bones.

17. The membranous areas (soft spots) of an infant's skull are called _____ .

18. What part of the vertebral column acts as a shock absorber?
    a. vertebral bodies
    b. intervertebral disks
    c. lamina
    d. spinous processes

19. Which of the vertebrae support the most weight?
    a. cervical
    b. thoracic
    c. lumbar
    d. sacral

20. The functions of the thoracic cage include
    a. production of blood cells.
    b. contribution to breathing.
    c. protection of heart and lungs.
    d. support of the shoulder girdle.

21. True ribs articulate with _____ _____ and the
    _____ .

22. The pectoral girdle is made up of two _____ and two _____ .

23. The _____ crosses over the ulna when the palm of the hand faces backward.

24. The wrist consists of
    a. 8 carpal bones.
    b. 5 metacarpal bones.
    c. 14 phalanges.
    d. distal segments of the radius and the ulna.

25. When the hands are placed on the hips, they are placed over
    a. the iliac crest.
    b. the acetabulum.
    c. the ischial tuberosity.
    d. the ischial spines.

26. The longest bone in the body is the
    a. tibia.
    b. fibula.
    c. femur.
    d. patella.

27. The lower end of the fibula can be felt as an ankle bone. The correct name is the
    a. head of the fibula.
    b. lateral malleolus.
    c. talus.
    d. lesser trochanter.

28. Synovial membrane is found in
    a. immovable joints.
    b. slightly movable joints.
    c. freely movable joints.

29. The function of bursae is
    a. to act as shock absorbers.
    b. to facilitate movement of tendon over bone.
    c. to reduce friction between bony surfaces.
    d. to protect joints from infection.

30. The type of joint that permits the widest range of motion is
    a. ball-and-socket.
    b. gliding.
    c. condyloid.
    d. pivot.

31. Moving the parts at a joint so that the angle between them is increased is called
    a. flexion.
    b. extension.
    c. elevation.
    d. abduction.

Study Activities

I. Aids to Understanding Words

Define the following word parts. (p. 127)

acetabul-	fov-
ax-	glen-
-blast	hema-
carp-	inter-
-clast	intra-
condyl-	meat-
corac-	odont-
cribr-	poie-
crist-	

II. Bone Structure (pp. 128–29)

A. Label the following parts in the accompanying drawing of a long bone (see opposite page): epiphysis, diaphysis, articular cartilage, spongy bone, red marrow, compact bone, medullary cavity, yellow marrow, periosteum. (p. 128)

B. What is the structural difference between compact and spongy bone? (p. 128)

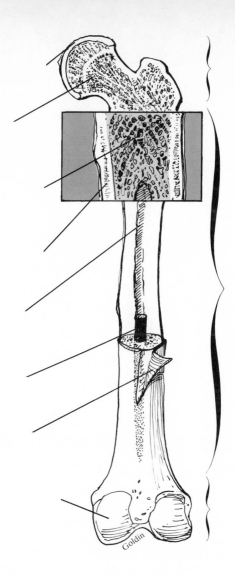

Goldin

III. Bone Development and Growth (pp. 130–31)

A. What bones are intramembranous bones? How do these develop? (p. 130)

B. What bones are endochondral bones? How do these develop? Be sure to include descriptions of the primary ossification center, the secondary ossification center, and the epiphyseal disk. (p. 130)

C. Answer these questions concerning ossification. (pp. 130–31)
   1. When is ossification complete?

   2. How can an X ray show that long bone growth is complete?

IV. Functions of Bones (pp. 131–33)

    A.   What bones function primarily to provide support? (p. 131)

    B.   What bones function primarily to protect viscera? (p. 131)

    C.   Answer these questions concerning blood cell formation. (pp. 132–33)
        1.   Where are blood cells formed in the embryo? In the infant? In the adult?

        2.   What is the difference between red and yellow marrow?

    D.   Answer these questions concerning the inorganic compounds in bone. (p. 133)
        1.   What are the major inorganic salts stored in bone? What other salts and heavy metals can also be stored in bone?

        2.   How is calcium released from bone so that it is available for physiologic processes?

        3.   What is osteoporosis?

V. Organization of the Skeleton (pp. 134–35)

    A.   What are the two major divisions of the skeleton? (p. 134)

    B.   List the bones found in each of these major divisions. (pp. 134–35)

VI. The Skull (pp. 135–41)

    A.   Answer these questions concerning the number of bones in the skull. (p. 136)
        1.   How many bones are found in the human skull?
        2.   How many of these bones are found in the cranium?
        3.   How many are found in the facial skeleton?

    B.   Answer these questions concerning the cranial bones. (pp. 136–40)
        1.   Using your own head or that of a partner, locate the following cranial bones and identify the suture lines that form their boundaries: occipital bone, temporal bones, frontal bones, and parietal bones.
        2.   What are the remaining two bones of the cranium? Where are they located?

    C.   Answer these questions concerning the facial bones. (pp. 140–41)
        1.   Using yourself or a partner, locate the following facial bones: maxilla, palatine, zygomatic, lacrimal bones, nasal bones, vomer, inferior nasal conchae, and mandible.
        2.   Which of the facial bones is known as the keystone of the face? Why?

        3.   Which of the facial bones is the only movable bone of the skull?

        4.   Describe the differences between the infant and the adult skull.

VII. The Vertebral Column (pp. 142–44)

A. What is the function of the vertebral column? Of intervertebral disks? (p. 142)

B. Label the following parts of the accompanying diagram: lamina, spinous process, transverse process, pedicle, vertebral foramen, superior articulating process, body. (p. 142)

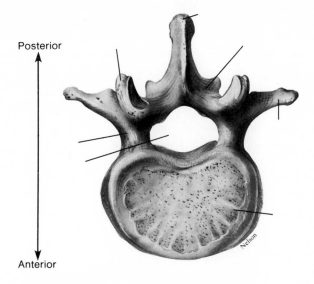

Posterior

Anterior

C. In what ways is the structure of the thoracic vertebrae unique? (p. 143)

D. In what ways is the structure of the lumbar vertebrae unique? (p. 144)

E. Locate the sacrum and the coccyx. (p. 144)

VIII. The Thoracic Cage (pp. 144–46)

A. Name the bones of the thoracic cage. (p. 144)

B. Describe the differences between true and false ribs. (p. 145)

C. Describe the sternum, including the manubrium, body, and xiphoid process. Locate these structures on yourself. (pp. 145–46)

IX. The Pectoral Girdle (p. 146)

Using yourself or a partner, locate and list the bones of the pectoral girdle. What is the function of the pectoral girdle?

X. The Upper Limb (pp. 147–49)

A. Using yourself or a partner, locate and list the bones of the upper limb. (p. 147)

B. Label the following parts in the drawing below: phalanges, metacarpals, carpals, pisiform, triangular, hamate, lunate, capitate, scaphoid, trapezoid, trapezium. (p. 149)

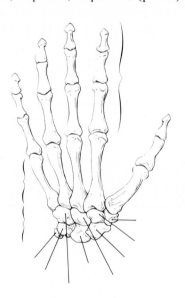

XI. The Pelvic Girdle (pp. 149–50)

A. List the bones of the pelvic girdle. (p. 149)

B. Identify the bone in which each of the following structures is located and explain the function of each structure. (pp. 149–50)

Structure and function of bones of the pelvic girdle

Structure	Bone	Function
Acetabulum		
Anterior superior iliac spine		
Symphysis pubis		
Obturator foramen		
Ischial tuberosity		
Ischial spine		

XII. The Lower Limb (pp. 151–53)

A. List the bones of the lower limb. (p. 151)

B. Identify the bone in which each of the following structures is located and explain the function of each structure. (p. 153)

Structure and function of bones of the lower limb

Structure	Bone	Function
Fovea capitis		
Medial malleolus		
Lateral malleolus		
Greater and lesser trochanters		
Tibial tuberosity		

C. Label these structures on the accompanying illustration: tarsal bones, calcaneus, talus, metatarsals, phalanges, navicular, cuboid, lateral cuneiform, intermediate cuneiform, medial cuneiform, proximal phalanx, middle phalanx, distal phalanx. (p. 153)

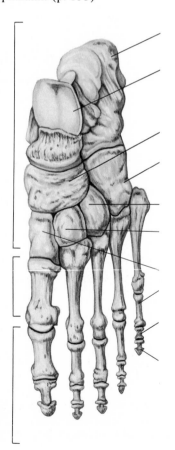

XIII. The Joints (pp. 154–59)

A. Describe and give an example of each of the following types of joints. (pp. 154–56)

The joints

Joint	Description	Example
Immovable joint		
Slightly movable joint		
Freely movable joint		
Ball-and-socket joint		
Condyloid joint		
Gliding joint		
Hinge joint		
Pivot joint		
Saddle joint		

B. Describe the structure and function of a freely movable joint. (pp. 154-55)

C. Identify the following movements. (pp. 156–59)
turning the palms of the hands upward
shrugging the shoulders
bending the arm at the elbow
reaching for an object that is just beyond one's reach
turning the hands down so the palms face the floor
moving the legs apart in an "at ease" position
moving the legs together in an "attention" position
swiveling the head
drawing a large circle on the blackboard
pointing the toes together with the heels apart

When you have completed the study activities to your satisfaction, retake the mastery test and compare your performance with your initial attempt. If there are still areas you do not understand, repeat the appropriate study activities.

8 Muscular System

Overview

This chapter presents the muscular system. In conjunction with the skeletal system, the muscular system serves to move the body. The chapter introduces the three types of muscle; the major events in contraction of skeletal, smooth, and cardiac muscles; the energy supply to muscle fiber for contraction; the occurrence of oxygen debt; and the process of muscle fatigue (objectives 1, 3, 4, 5, and 10). The chapter describes the structure and function of a skeletal muscle; distinguishes between a twitch and a sustained contraction; explains how various kinds of muscle contraction produce body movements and maintain posture; shows how the location and interaction of muscles produce body movements; identifies the location and action of major skeletal muscles; and differentiates the structure and function of a multiunit smooth muscle and a visceral smooth muscle (objectives 2, 6, 7, 8, 9, 11, and 12).

The skeletal system can be thought of as the passive partner in producing movement; the muscular system can be thought of as the active partner. This chapter explains how muscles interact with bones to maintain posture and produce movement. In addition, it tells the characteristics and functions of skeletal, smooth, and cardiac muscles. This knowledge is a foundation for the study of other organ systems, such as the digestive system, the respiratory system, and the cardiovascular system.

Chapter Objectives

After you have studied this chapter, you should be able to

1. Describe how connective tissue is included in the structure of a skeletal muscle.
2. Name the major parts of a skeletal muscle fiber and describe the function of each part.
3. Explain the major events that occur during muscle fiber contraction.
4. Explain how energy is supplied to the muscle fiber contraction mechanism.
5. Describe how oxygen debt develops and how a muscle may become fatigued.
6. Distinguish between a twitch and a sustained contraction.
7. Explain how various types of muscular contractions are used to produce body movements and maintain posture.
8. Describe how skeletal muscles are affected by exercise.
9. Distinguish between the structure and function of a multiunit smooth muscle and a visceral smooth muscle.
10. Compare the fiber contraction mechanisms of skeletal, smooth, and cardiac muscles.
11. Explain how the locations of skeletal muscles are related to the movements they produce and how muscles interact in producing such movements.
12. Identify and describe the location of the major skeletal muscles of each body region and describe the action of each muscle.

Focus Question

How do muscle cells use energy and interact with bones to accomplish movement?

Mastery Test

Now take the mastery test. Do not guess. As soon as you complete the test, correct it. Note your successes and failures so that you can read the chapter to meet your learning needs.

1. An individual muscle is separated from adjacent muscles by _____ .

2. Striated muscle tissue is also known as
   a. voluntary muscle.
   b. involuntary muscle.
   c. skeletal muscle.
   d. visceral muscle.

3. Layers of connective tissue extending into the muscle to form partitions between muscle bundles are continuous with attachments of muscle to periosteum called
   a. ligaments.
   b. tendons.
   c. aponeuroses.
   d. elastin.

4. The characteristic striated appearance of skeletal muscle is due to the arrangement of alternating protein filaments composed of _____ and _____ .

5. The union between a nerve fiber and a muscle fiber is the
   a. motor neuron.
   b. motor end plate.
   c. neuromuscular junction.
   d. neurotransmitter.

6. When cross-bridges form between filaments of actin and myosin, the result is
   a. shortening of the muscle fiber.
   b. membrane polarization.
   c. release of acetylcholine.

7. The substance that halts the stimulation of a muscle fiber by a neuron is _____ .

8. The energy used in muscle contraction is supplied by the decomposition of _____
   _____ .

9. A substance that stores energy released when stores of the substance in number 8 are in low supply is
   _____ _____ .

10. A person feels out of breath after vigorous exercise because of oxygen debt. Which of the following statements helps explain this phenomenon?
    a. Anaerobic respiration increases during strenuous activity.
    b. Lactic acid is metabolized more efficiently when the body is at rest.
    c. Conversion of lactic acid to glycogen occurs in the liver and requires energy.
    d. Priority in energy use is given to ATP synthesis.

11. After prolonged muscle use, muscle fatigue occurs due to an accumulation of _____
    _____ .

12. The minimal strength stimulus needed to elicit contraction of a single muscle fiber is called
    a(n) _____ _____ .

13. The strength of a muscle contraction in response to different levels of stimulation is determined by
    a. the level of stimulation delivered to individual muscle fibers.
    b. the number of fibers that respond in each motor unit.
    c. the number of motor units stimulated.
    d. the characteristics of each muscle group.

14. The period of time between a stimulus to a muscle and muscle response is called the
    a. latent period.
    b. contraction.
    c. refractory period.

15. Muscle tone refers to
    a. a state of sustained, partial contraction of muscles that is necessary to maintain posture.
    b. a feeling of well-being following exercise.
    c. the ability of a muscle to maintain contraction against an outside force.
    d. the condition athletes attain after intensive training.

16. Smooth muscle contracts (more slowly, more rapidly) than skeletal muscle following stimulation.

17. Two types of smooth muscle are _____ muscle and _____ muscle.

18. Peristalsis is due to which of the following characteristics of smooth muscle?
    a. capacity of smooth muscle fibers to excite each   c. rhythmicity
       other   d. sympathetic innervation
    b. automaticity

19. Impulses travel relatively (rapidly, slowly) through cardiac muscle.

20. The attachment of a muscle to a relatively fixed part is called the _____ ; the attachment to a relatively movable part is called the _____ .

21. Smooth body movements depend on _____ giving way to prime movers.

22. The muscle that compresses the cheeks inward when it contracts is the
    a. orbicularis oris.   c. platysma.
    b. epicranius.   d. buccinator.

23. The muscle that moves the head to one side is the
    a. sternocleidomastoid.   c. semispinalis capitis.
    b. splenius capitis.   d. longissimus capitis.

24. The muscle that abducts the upper arm and can both flex and extend the humerus is the
    a. biceps brachii.   c. infraspinatus.
    b. deltoid.   d. triceps brachii.

25. The band of tough connective tissue that extends from the xiphoid process to the symphysis pubis and serves as an attachment for muscles of the abdominal wall is the _____

    _____ .

26. The heaviest muscle in the body, which serves to straighten the leg at the hip during walking, is the
    a. psoas major.   c. adductor longus.
    b. gluteus maximus.   d. gracilis.

Study Activities

I. Aids to Understanding Words

Define the following word parts. (p. 167)

calat-   myo-

erg-   syn-

hyper-   tetan-

inter-   -troph

laten-

II. Structure of a Skeletal Muscle (pp. 168–71)

A. List the kinds of tissue present in skeletal muscle. (p. 168)

B. A skeletal muscle is held in position by layers of fibrous connective tissue called _____ .
    This tissue extends beyond the end of a skeletal muscle to form a cordlike _____ .
    When this tissue extends beyond the muscle to form a sheetlike structure, it is called a(n)

    _____ .
    What are fascicles? _____ (p. 168)

C. Answer these questions concerning skeletal muscle fibers. (pp. 169–70)
    1. The contractile unit of a muscle is a(n) _____ _____ _____ .
    2. Describe the structure and function of a sarcomere.

    3. The network of membranous channels in the cytoplasm of muscle fibers is the
        _____ _____ .

    4. What is the function of these channels?

D. Describe the transmission of a nerve impulse across the neuromuscular junction. (pp. 170–71)

E. Label these structures in the accompanying illustration of a single fiber motor unit: mitochondria, synaptic cleft, synaptic vesicles, folded sarcolemma, muscle fiber nucleus, myofibril of muscle fiber, motor end plate, nerve fiber branches, motor neuron fiber. (p. 171)

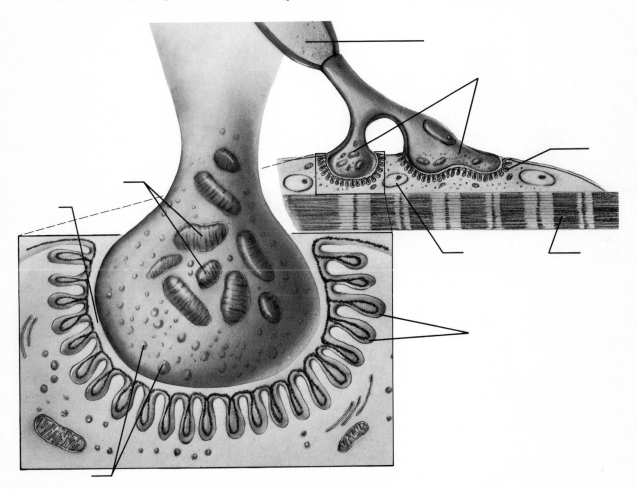

III. Skeletal Muscle Contraction (pp. 171–75)

A.   Describe the roles of actin and myosin in muscle contraction. (p. 171)

B.   Answer these questions concerning stimulus for contraction. (pp. 171–72)
1.   Describe the interaction of acetylcholine and calcium ions in stimulating muscle contraction.

2.   The action of acetylcholine is halted by the enzyme _____ .

C.   Answer these questions concerning energy sources for contraction. (p. 173)
1.   How does ATP supply energy for muscle contraction?

2.   How does creatine phosphate supply energy for muscle contraction?

D.   Answer these questions concerning oxygen supply and cellular respiration. (p. 174)
1.   What substance in muscle seems able to store oxygen temporarily?

2.   Why is oxygen necessary for muscle contraction?

3.   How does the muscle continue to contract in the absence of oxygen?

4.   What is meant by oxygen debt? How is it paid off?

E.   Answer these questions concerning muscle fatigue. (p. 175)
1.   What is meant by muscle fatigue? What causes it?

2.   About 25% of the energy released by cellular respiration is available for metabolic processes. About 75% is lost as _____ .

IV. Muscular Responses (pp. 175–78)

A.   Answer these questions concerning muscular responses. (p. 175)
1.   Define *threshold stimulus*.

2.   Define *all-or-none response*.

3.   How is all-or-none response related to the strength of a muscle contraction?

4.   Describe recruitment.

B. The accompanying illustration shows a myogram of a type of muscle contraction known as a twitch. Label the following: time of stimulation, latent period, period of contraction, period of relaxation. Explain the significance of each of these events. (p. 176)

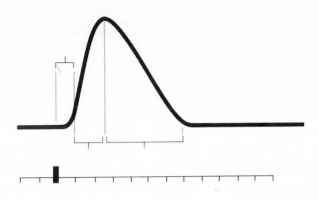

C. Describe these kinds of muscle contractions. (p. 176)

sustained contraction

tetanic contraction

muscle tone

D. Complete the following chart. (p. 177)

Changes in muscle size		
Change	Conditions Leading to Change	Observable Change in Muscle
Hypertrophy		
Atrophy		

V. Smooth and Cardiac Muscles (p. 178)

Complete the following chart.

Types of muscle tissue	Skeletal	Smooth	Cardiac
Major location			
Major function			
Mode of control			
Contraction characteristics			

VI. Skeletal Muscle Actions (pp. 178–80)

A.   A skeletal muscle has at least two places of attachment to bone. For instance, the gluteus maximus, which extends the leg at the hip, is attached to the posterior surface of the ilium, the sacrum, and the coccyx at one end and to the posterior surface of the femur and the iliotibial tract at the other. One place of muscle attachment is the origin and the other is the insertion. Explain the difference between the two. (p. 180)

B.   Match the terms in the first column with the statements in the second column that best describe the role of muscle groups in producing smooth muscle movement. (p. 179)

____ 1. prime mover
____ 2. synergist
____ 3. antagonist

a. muscle that returns a part to its original position
b. muscle that makes the action of the prime mover more effective
c. muscle that has the major responsibility for producing a movement

When you have completed the study activities to your satisfaction, retake the mastery test and compare your performance with your initial attempt. If there are still areas you do not understand, repeat the appropriate study activities.

9 Nervous System

Overview

The human body uses two systems to coordinate and integrate the functions of other body systems so that the internal environment remains stable. These systems are the nervous system and the endocrine system. This chapter focuses on the nervous system. It explains the structure and function of the various parts of the nervous system; the neurons, the brain and spinal cord and their coverings; the spinal nerves; and a reflex arc (objectives 1, 7–10, and 15). It describes the organization of the nervous system; neurons and their classification; the organization of the brain; the parts of the peripheral and autonomic nervous systems; and the cranial nerves and their functions (objectives 6, 11, 13, 14, 16, 17, and 18). It tells how the nervous system transmits messages (objectives 3, 4, and 5) and how the system protects itself from injury (objective 12). It also describes the structure and function of the various connective tissues in the nervous system (objective 2).

The maintenance of the internal environment within rather narrow limits is called homeostasis and is a necessary condition of life. Knowledge of the nervous system and its functions is basic to understanding how this is accomplished.

Chapter Objectives

After you have studied this chapter, you should be able to

1. Describe the general structure of a neuron.
2. Name four types of neuroglial cells and describe the functions of each.
3. Describe the events that lead to the conduction of a nerve impulse.
4. Explain how a nerve impulse is transmitted from one neuron to another.
5. Explain two ways impulses are processed in neuronal pools.
6. Explain how differences in structure and function are used to classify neurons.
7. Name the parts of a reflex arc and describe the function of each part.
8. Describe the coverings of the brain and spinal cord.
9. Describe the structure of the spinal cord and its major functions.
10. Name the major parts of the brain and describe the functions of each part.
11. Distinguish between motor, sensory, and association areas of the cerebral cortex.
12. Describe the formation and the function of cerebrospinal fluid.
13. List the major parts of the peripheral nervous system.
14. Name the cranial nerves and list their major functions.
15. Describe the structure of a spinal nerve.
16. Describe the functions of the autonomic nervous system.
17. Distinguish between the sympathetic and the parasympathetic divisions of the autonomic nervous system.
18. Describe a sympathetic and a parasympathetic nerve pathway.

Focus Question

It is noon and you are just finishing an anatomy assignment. You hear your stomach growling. You realize you are hungry. You make a ham sandwich and pour a glass of milk. After you finish eating, you decide you have been studying for three hours and should go for a walk. How does the nervous system receive internal and external cues, process incoming information, and decide what action to take?

Mastery Test

Now take the mastery test. Do not guess. As soon as you complete the test, correct it. Note your successes and failures so that you can read the chapter to meet your learning needs.

1. The three general functions of the peripheral nervous system are _____ , _____ , and _____ .

2. The basic unit of structure and function of the nervous system is the _____ .

3. Which of the following structures is *not* common to all nerve cells?
   - a. cell body
   - b. axon
   - c. dendrite
   - d. myelin sheath

4. Does statement *a* explain statement *b?* _____
   - a. The nucleus of the nerve cell seems incapable of mitosis.
   - b. The nerve cell cannot reproduce.

5. The neurilemma is composed of
   - a. Nissl bodies.
   - b. myelin.
   - c. the cytoplasm and nuclei of Schwann cells.
   - d. neuron cell bodies.

6. All but one of the following are functions of the neuroglial cells.
   - a. fill spaces
   - b. support growth of neurons
   - c. hold organs together
   - d. phagocytize bacteria

7. Potassium/sodium ions tend to pass more easily through the cell membrane of nerve cells.

8. When the nerve cell is at rest the concentration of _____ ions is relatively greater on the outside of the cell membrane.

9. When the threshold potential is reached, the region of cell membrane being stimulated undergoes a change in _____ .

10. In which type of fiber is conduction faster?
    - a. myelinated
    - b. unmyelinated

11. The junction of two neurons is called a(n) _____ .

12. The function of neuronal pools is _____ of nerve impulses.

13. Continuous stimulation of a neuron on the distal side of this junction is prevented by
    - a. exhaustion of the nerve fiber.
    - b. the chemical instability of neurotransmitters.
    - c. enzymes within the neural junction.
    - d. rapid depletion of ionized calcium.

14. The most common neuron structure is
    - a. one axon and many dendrites.
    - b. one process that serves as both axon and dendrite.
    - c. one dendrite and many axons.
    - d. one dendrite and one axon.

15. Neurons may be classified functionally as _____ , _____ , and _____ neurons.

16. A bundle of nerve fibers held together by connective tissue is a(n) _____ .

17. An automatic, unconscious response to a change inside or outside the body is a(n) _____ .

18. The organs of the central nervous system are the _____ and the _____ .

19. The outer membrane covering the brain is composed of fibrous connective tissues and is called the
    a. dura mater.   c. pia mater.
    b. arachnoid mater.   d. periosteum.

20. Cerebrospinal fluid is found between
    a. the arachnoid mater and the dura mater.   c. the pia mater and the arachnoid mater.
    b. the vertebrae and the meninges.

21. The spinal cord ends
    a. at the sacrum.   c. between lumbar vertebrae 1 and 2.
    b. between thoracic vertebrae 11 and 12.   d. at lumbar vertebra 5.

22. Which of the following statements is(are) true about the white matter in the spinal cord?
    a. A cross section of the cord reveals a core of   c. The white matter carries sensory stimuli to the
       white matter surrounded by gray matter.   brain; the gray matter carries motor stimuli to the
    b. The white matter is composed of myelinated   periphery.
       nerve fibers and makes up nerve pathways   d. The nerve fibers within spinal tracts arise from cell
       called tracts.   bodies located in the same part of the nervous
                                                            system.

23. The three major portions of the brain are the _____ , _____ , and
    _____ .

24. The hemispheres of the cerebrum are connected by nerve fibers called the
    a. corpus callosum.   c. tissue of Rolando.
    b. falx cerebri.   d. tentorium.

25. Match the functions in the first column with the appropriate area of the brain in the second column.
    ____ a. hearing   1. frontal lobes
    ____ b. vision   2. parietal lobes
    ____ c. recognition of printed work   3. temporal lobes
    ____ d. control of voluntary muscles   4. occipital lobes
    ____ e. pain
    ____ f. complex problem solving

26. Which hemisphere of the brain is dominant for most of the population?

27. Cerebrospinal fluid is produced by the _____ _____ .

28. The thalamus and hypothalamus are parts of the brain located in the
    a. midbrain.   c. medulla oblongata.
    b. pons.   d. diencephalon.

29. The part of the brain responsible for regulation of temperature and heart rate, control of hunger, and regulation
    of fluid and electrolytes is the
    a. thalamus.   c. medulla oblongata.
    b. hypothalamus.   d. pons.

30. The _____ _____ produces emotional reactions of fear, anger, and
    pleasure.

31. Consciousness is dependent on stimulation of the _____ _____ .

32. The peripheral nervous system has two divisions, the _____ nervous system and the _____ nervous system.

33. There are _____ pairs of cranial nerves; all but one of these arise from the

_____ _____ .

34. There are _____ pairs of spinal nerves.

35. The part of the nervous system that functions without conscious control is the _____ nervous system.

36. Nerves of the sympathetic division leave the spinal cord with spinal nerves in the _____ and _____ .

37. Which of the following are responses to stimulation by the sympathetic nervous system?
   a. increased heart rate
   b. increased blood glucose concentration
   c. increased peristalsis
   d. increased salivation

38. Which of the following are responses to stimulation of the parasympathetic nervous system?
   a. dilation of the bronchioles
   b. dilation of the coronary arteries
   c. contraction of the gallbladder
   d. contraction of the muscles of the urinary bladder

Study Activities

I. Aids to Understanding Words

Define the following word parts. (p. 205)

ax-	moto-
dendr-	peri-
funi-	plex-
gangli-	sens-
-lemm	syn-
mening-	ventr-

II. Introduction, General Functions of the Nervous System, and Nerve Tissue (pp. 206–10)

A. List the organs of the central and peripheral nervous systems. (p. 206)

B. Describe the general functions of the nervous system. (p. 206)

C. Answer these questions concerning neuron structure. (pp. 206–8)

    1. Label the following structures in the drawing of a motor neuron below: dendrites, axon, nucleolus, cell body, nucleus, neurofibrils, Schwann cell, nodes of Ranvier, myelin.

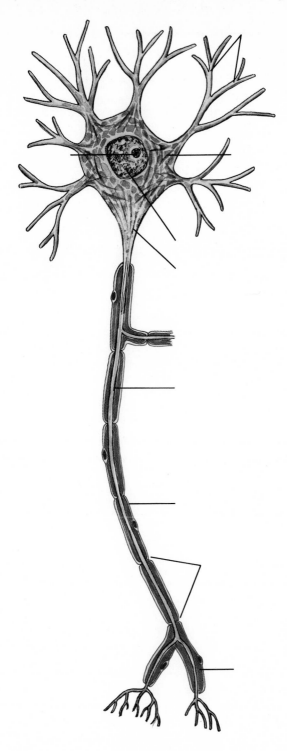

(a)

2. Describe how Schwann cells make up the myelin sheath and the neurilemma on the outsides of nerve fibers.

D. Fill in the following chart. (pp. 208–10)

Neuroglial cells

Cell	Location	Structure	Function
Astrocytes			
Microglia			
Oligodendrocytes			
Ependyma			

III. Cell Membrane Potential (pp. 210–12)

A. The outside of a cell membrane is usually electrically charged with respect to the inside, due to a(n) _____ distribution of _____ on either side of the membrane. (p. 210)

B. Describe the events occurring in the cell membrane that permit conduction of an impulse. (pp. 210–12)

IV. Nerve Impulse (p. 212)

A. Describe membrane polarization, depolarization, and repolarization. Which of these events is a nerve impulse? (p. 212)

B. Answer these questions concerning impulse conduction. (p. 212)
   1. How do the nodes of Ranvier affect nerve impulse conduction? What kind of conduction is this called?

   2. Define the all-or-none response in neurons.

V. Synapse and Processing of Impulses (pp. 213–16)

A. Answer these questions concerning synaptic transmission. (p. 214)

    1. Label the following structures in the accompanying drawing of a synapse: axon, mitochondrion, synaptic vesicles, synaptic cleft, dendrite, axon membrane, neurotransmitter substance, polarized membrane.

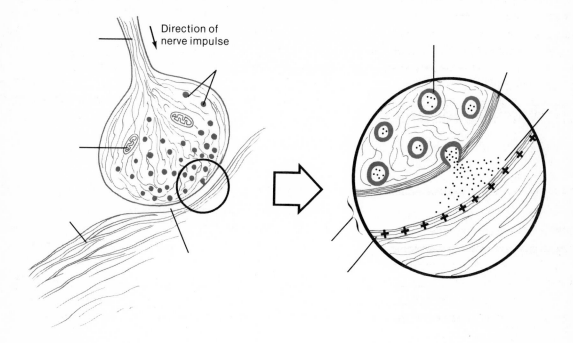

    2. How does a neurotransmitter initiate depolarization?

    3. How is stimulation of a nerve fiber stopped?

    4. List substances that function as neurotransmitters.

B. Answer these questions concerning excitatory and inhibitory actions. (p. 215)

    1. Describe excitatory and inhibitory actions. How do they interact in normal nerve function?

    2. What substances seem to have inhibitory action?

C. Describe the role of neuronal pools in producing facilitation, convergence, and divergence. (pp. 215–16)

VI. Types of Neurons and Nerves (p. 217)

A. Neurons can be classified by structure. Describe and locate these neurons. (p. 217)

multipolar neurons

bipolar neurons

unipolar neurons

B. Neurons can also be classified by function. Fill in the following chart. (p. 217)

Location and function of neurons classified by function

Neuron	Location	Function
Sensory neurons		
Interneurons		
Motor neurons		

C. What is a nerve? What is a motor nerve; a sensory nerve; a mixed nerve? (p. 217)

VII. Nerve Pathways (pp. 217–19)

A. What is a reflex? What is a reflex arc? (p. 218)

B. Label the following parts of the reflex shown in the drawing: dendrite of sensory neuron, cell body of sensory neuron, axon of sensory neuron, dendrite of motor neuron, cell body of motor neuron, axon of motor neuron. (p. 218)

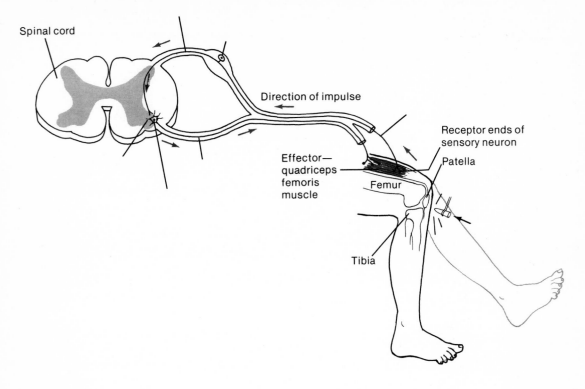

Spinal cord

Direction of impulse

Receptor ends of sensory neuron

Effector—quadriceps femoris muscle

Patella

Femur

Tibia

VIII. Coverings of the Central Nervous System (p. 220)

A. What are the bony coverings of the central nervous system? (p. 220)

B. Fill in the following chart. (p. 220)

The meninges

Layer	Location	Structure and Special Features	Function
Dura mater			
Arachnoid mater			
Pia mater			

IX. Spinal Cord (pp. 221–23)

A. Answer these questions concerning the structure of the spinal cord. (pp. 221–22)

1. The superior boundary of the spinal cord is _____ . The inferior boundary of the spinal cord is _____ .

2. Label the accompanying drawing of a cross section of the spinal cord, showing the anterior median fissure, posterior median sulcus, white matter, gray matter, posterior horn, lateral horn, anterior horn, gray commissure, and central canal.

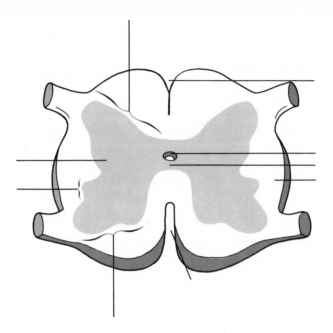

B. What is the effect of injury to ascending tracts? To descending tracts? (p. 222)

X. Brain (pp. 223–31)

A. Answer these questions concerning the structure of the cerebrum. (pp. 223–25)

1. List the lobes of the cerebral hemispheres.

2. The bridge that connects the two hemispheres is the _____ _____ .

3. The ridges of the hemispheres are _____ .

4. A shallow groove is a(n) _____ ; a deeper groove is a(n) _____ .

5. The outer layer of the cerebrum is the _____ . It is composed of _____ _____ .

6. The inner layer of the cerebrum is composed of _____ _____ .

7. Masses of gray matter deep within the cerebrum that inhibit motor activity are the _____ _____ .

B. Answer these questions concerning the function of the cerebrum. (pp. 226–28)
    1. What are the functions of the cerebrum?

    2. Fill in the following chart.

Functional areas of the cerebrum

Areas	Locations	Function
Motor		
Sensory		
Association		

C. Describe the functions of the dominant and nondominant hemispheres of the brain. (p. 228)

D. Answer these questions concerning the ventricles and cerebrospinal fluid. (pp. 228–29)
    1. Describe the location of the four ventricles.

    2. Where is cerebrospinal fluid secreted? What is its function?

E. Answer these questions concerning the brain stem. (pp. 230–31)
    1. Where is the brain stem? What are its component structures?

    2. Locate the diencephalon and describe its structure; then fill in the following chart.

Functions of the diencephalon

Structure	Location	Function
Thalamus		
Hypothalamus		

3.   What structures make up the limbic system? What is the function of this system?

4.   Where is the midbrain located? What is its function?

5.   Where is the pons located? What is its function?

6.   Where is the medulla oblongata located? The medulla is the control center for what vital activities?

7.   Describe the location, structure, and function of the reticular formation.

F.   What is the function of the cerebellum? (p. 231)

XI. Peripheral Nervous System and Autonomic Nervous System (pp. 232–39)

A.   Answer these questions concerning the parts of the peripheral nervous system. (p. 232)
  1.   What are the parts of the peripheral nervous system?

  2.   What is the function of the somatic nervous system? The autonomic nervous system?

B.   Answer these questions concerning the cranial nerves. (pp. 232–33)
  1.   An easy way to memorize the names of the cranial nerves is to use the following sentence.

On old Olympus towering tops a   Finn and   German viewed some hops.
I   II III   IV   V   VI VII VIII IX   X   XI XII

(Remember that VIII can be called auditory, and XI can be called spinal accessory.)
You need to be able to identify the cranial nerves by both name and number.

2. Fill in the following chart.

Cranial nerve functions

Cranial Nerve	Sensory, Motor, or Mixed	Function
I Olfactory		
II Optic		
III Oculomotor		
IV Trochlear		
V Trigeminal		
VI Abducens		
VII Facial		
VIII Vestibulocochlear or Auditory		
IX Glossopharyngeal		
X Vagus		
XI Accessory		
XII Hypoglossal		

C. Answer these questions concerning spinal nerves. (pp. 233–35)
   1. How are the spinal nerves identified?

   2. What is the structure and function of the dorsal root? Of the ventral root?

D. Answer these questions concerning spinal nerve plexuses. (p. 235)
   1. What is a plexus?

   2. Fill in the following chart.

Spinal nerve plexuses

	Nerves Involved	Structures Innervated
Cervical plexuses		
Brachial plexuses		
Lumbosacral plexuses		

E. Answer these questions concerning the autonomic nervous system. (pp. 235–39)
   1. What structures make up the autonomic nervous system, and what is the function of this system?

   2. How are the nerve pathways of the autonomic division different from those of the somatic division?

When you have completed the study activities to your satisfaction, retake the mastery test and compare your performance with your initial attempt. If there are still some areas you do not understand, repeat the appropriate study activities.

10 Somatic and Special Senses

Overview

This chapter deals with specialized parts of the nervous system that allow the body to assess and adjust to the external environment. It describes the locations and structures of the somatic and special senses and the function of each of them in maintaining homeostasis (objectives 1–11).

An understanding of these senses is necessary to knowing how the nervous system receives input and responds to support life.

Chapter Objectives

After you have studied this chapter, you should be able to

1. Name five kinds of receptors and explain the function of each kind.
2. Explain how a sensation is produced.
3. Describe the somatic senses.
4. Describe the receptors associated with the senses of touch and pressure, temperature, and pain.
5. Describe how the sense of pain is produced.
6. Explain the relationship between the senses of smell and taste.
7. Name the parts of the ear and explain the function of each part.
8. Distinguish between static and dynamic equilibrium.
9. Name the parts of the eye and explain the function of each part.
10. Explain how light is refracted by the eye.
11. Describe the visual nerve pathway.

Focus Question

When you began this chapter at 3:00 P.M., it was 32° outside, but the sun was pouring into the room. It is now after 5:00, and as you reach to turn on the light, you notice the room has become chilly, so you get a sweater. You smell the supper your roommate is preparing, and you realize you are hungry. How have your special senses functioned to process and act on this sensory information?

Mastery Test

Now take the mastery test. Do not guess. As soon as you complete the test, correct it. Note your successes and failures so that you can read the chapter to meet your learning needs.

1. Sensory receptors are sensitive to stimulation by
   a. changes in concentration of chemicals.
   b. temperature changes.
   c. tissue damage.
   d. mechanical forces.
   e. changes in intensity of light.

2. Sensory receptors for all of the following adapt to repeated stimulation by sending fewer and fewer impulses, *except* those for
   a. heat.
   b. light.
   c. pain.
   d. touch.

3. Meissner's corpuscles and Pacinian corpuscles are sensitive to
   a. touch and pressure.
   b. pain.
   c. heat.
   d. light.

4. Pain receptors respond to the release of _____ .

5. Which of the following events will elicit pain from visceral organs?
   a. spasm of smooth muscle
   b. cutting into the viscera
   c. stretching of a visceral organ
   d. burning, as in electrocautery

6. Pain from the heart is likely to be experienced in the left shoulder. This is an example of _____ pain.

7. The receptors for taste and smell are examples of
   a. mechanical receptors.
   b. chemoreceptors.
   c. thermoreceptors.

8. Olfactory receptors are located in
   a. the nasopharynx.
   b. the inferior nasal conchae.
   c. the superior nasal conchae.
   d. the lateral wall of the nostril.

9. Impulses that stimulate the olfactory receptors are transmitted along the _____

   _____ .

10. The sensitive part of a taste bud is the
    a. taste cell.
    b. taste pore.
    c. taste hair.

11. Saliva enhances the taste of food by
    a. increasing the motility of taste receptors.
    b. dissolving the chemicals that cause taste.
    c. releasing taste factors by partially digesting food.

12. The four primary taste sensations are _____ , _____ ,

    _____ , and _____ .

13. In addition to the sense of hearing, the ear also functions in the sense of _____ .

14. The functions of the small bones of the middle ear are to
    a. provide a framework for the tympanic membrane.
    b. protect the structures of the inner ear.
    c. transmit vibrations from the external ear to the inner ear.
    d. increase the force of vibrations transmitted to the inner ear.

15. A means of providing equal pressure on both sides of the eardrum is furnished by the

    _____ _____ .

16. The inner ear consists of two complex structures called the _____ _____
    and the _____ _____ .

17. Sound is transmitted in the inner ear via a fluid called _____ .

18. Hearing receptors are located in the
    a. organ of Corti.
    b. scala vestibuli.
    c. scala tympani.
    d. round window.

19. The hair cells of the inner ear are stimulated by
    a. bending the head forward or backward.
    b. rapid turns of the head or body.
    c. changes in the position of the body relative to the ground.
    d. changes in the position of skeletal muscles.

20. The muscle that raises the eyelid is the
    a. orbicularis oculi.
    b. superior rectus.
    c. levator palpebrae superioris.
    d. ciliary muscle.

21. The conjunctiva covers the anterior surface of the eyeball, except for the _____ .

22. The superior rectus muscle rotates the eye
    a. upward and toward the midline.
    b. toward the midline.
    c. away from the midline.
    d. upward and away from the midline.

23. The transparency of the cornea is due to
    a. the nature of the cytoplasm in the cells of the cornea.
    b. the small number of cells and the lack of blood vessels.
    c. the lack of nuclei with these cells.
    d. keratinization of cells in the cornea.

24. In the posterior wall of the eyeball, the sclera is pierced by the _____ _____ .

25. The shape of the lens changes as the eye focuses on a close object in a process known as
    a. accommodation.
    b. refraction.
    c. reflection.
    d. strabismus.

26. The anterior chamber of the eye extends from the _____ to the iris.

27. The part of the eye that controls the amount of light entering the eye is the _____ .

28. The inner tunic of the eye contains the receptor cells of sight and is called the _____ .

29. The region associated with the sharpest vision is the
    a. macula lutea.
    b. fovea centralis.
    c. optic disk.
    d. choroid coat.

30. The bending of light waves as they pass at an oblique angle from a medium of one optical density to a medium of another optical density is called _____ .

31. There are two types of visual receptors; one is called _____ , and the other is called
    _____ .

32. Match the type of vision in the first column with the proper receptor from the second column.
    ____ a. vision in relatively dim light   1. rods
    ____ b. color vision   2. cones
    ____ c. general outlines
    ____ d. sharp images

33. The light-sensitive pigment in rods is _____ . In the presence of light, this pigment
    decomposes to form _____ and _____ .

34. With the eyes closed, a person can accurately describe the positions of the limbs. Which of the following structures serve in this function?
    a. mechanoreceptors
    b. vestibule
    c. frontal lobe of the cerebrum
    d. cerebellum

Study Activities

I. Aids to Understanding Words

Define the following words and word parts. (p. 247)

choroid   olfact-
cochlea   scler-
iris   tympan-
labyrinth   vitre-
lacri-
macula

II. Receptors and Sensations (p. 248)

   A.   List five groups of sensory receptors. (p. 248)

   B.   The process that allows an individual to locate the region of stimulation is called
        _____ . (p. 248)

   C.   The process that makes a receptor ignore a continuous stimulus unless the strength of that stimulus is
        increased is _____ _____ . (p. 248)

III. Somatic Senses (pp. 248–51)

   A.   Fill in the following chart. (pp. 248–50)

Cutaneous receptors

Type	Function	Sensation
Free nerve endings (mechanoreceptors)		
Meissner's corpuscles (mechanoreceptors)		
Free nerve endings (temperature senses—heat)		
Free nerve endings (temperature senses—cold)		
Free nerve endings (pain receptors)		

   B.   Answer these questions concerning pain receptors. (pp. 249–51)
        1.   How do pain receptors differ from the other somatic senses?

        2.   What events trigger visceral pain?

        3.   What is referred pain?

        4.   Compare tension headaches and vascular headaches.

IV. Sense of Smell (pp. 252–53)

   A.   The sense of smell supplements the sense of _____ . (p. 252)

B. On the accompanying illustration, label the olfactory tract, olfactory bulb, cribriform plate, nasal cavity, and the olfactory area of the nasal cavity. (p. 252)

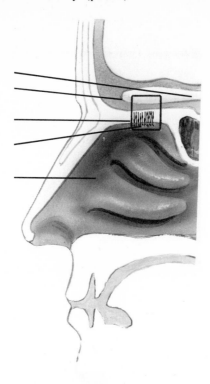

C. How do odors stimulate olfactory receptors? (p. 252)

V. Sense of Taste   (pp. 253–54)

A. Describe the structure of taste receptors. (p. 253)

B. How does saliva contribute to the perception of taste? (p. 253)

C. Identify the taste associated with the darkest areas in the accompanying illustration. (p. 254)

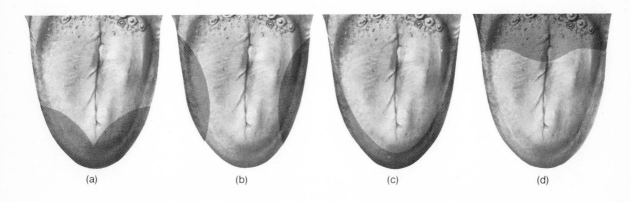

(a)   (b)   (c)   (d)

VI. Sense of Hearing (pp. 255–59)

A. Describe the function of the external ear. (p. 255)

B. Describe the vibration-conduction pathway of the ear from the meatus to the temporal lobe of the cerebrum. (pp. 255–59)

C. Why does it help to chew gum while descending in an airplane? (p. 258)

D. Describe the function of the inner ear. (pp. 256–59)

VII. Sense of Equilibrium   (pp. 260–61)

A. Distinguish between static and dynamic equilibrium. (pp. 260–61)

B. Describe the function of each of the following structures in maintaining equilibrium. (pp. 260–61)

utricle   crista ampullaris

saccule   cerebellum

macula   eyes

semicircular canals

VIII. Sense of Sight   (pp. 261–70)

A. Answer these questions concerning the visual accessory organs. (pp. 261–63)
1. What structures are covered by the conjunctiva?

2. Describe the lacrimal apparatus. How does it protect the eye?

3. Identify the function of the following muscles.
orbicularis oculi   medial rectus

levator palpebrae superioris   lateral rectus

superior rectus   superior oblique

inferior rectus   inferior oblique

B. Label the following structures in the accompanying illustration: cornea, lens, iris, suspensory ligaments, vitreous humor, aqueous humor, sclera, optic disk, optic nerve, fovea centralis, retina, choroid coat, pupil, ciliary body. (p. 264) What is the function of each of the labeled structures?

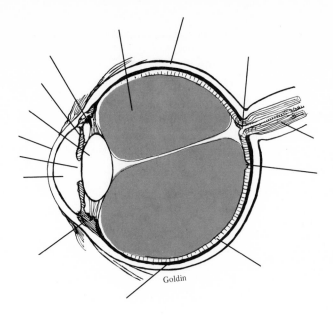

Goldin

C. Define *refraction of light*. (pp. 267–68)

D. Answer these questions concerning visual receptors. (pp. 268–69)
    1. Describe the functions of rods.

    2. Describe the location and functions of cones.

    3. What mechanism is used by cones to recognize color?

E. Describe the visual nerve pathway. (p. 270)

When you have completed the study activities to your satisfaction, retake the mastery test and compare your performance with your initial attempt. If there are still areas you do not understand, repeat the appropriate study activities.

11 Endocrine System

Overview

The endocrine system, like the nervous system, controls body activities to maintain a relatively constant internal environment. The methods used by these two systems are different. This chapter describes the difference between endocrine and exocrine glands, the location of the endocrine glands, and the hormones they secrete (objectives 1 and 5). It explains the nature of hormones; the substances that function as hormones; how hormones affect target tissues; how the secretion of hormones is controlled by a negative feedback system and the nervous system; the general function of each hormone; and the result of too little or too much of each hormone (objectives 2–4, 6, and 7).

A knowledge of the function of the endocrine system is basic to the understanding of how metabolic processes are regulated to meet the changing needs of the body.

Chapter Objectives

After you have studied this chapter, you should be able to

1. Distinguish between endocrine and exocrine glands.
2. Explain how steroid and nonsteroid hormones produce effects on target cells.
3. Discuss how hormone secretions are regulated by feedback mechanisms.
4. Explain how hormone secretions may be controlled by the nervous system.
5. Name and describe the location of the major endocrine glands and list the hormones they secrete.
6. Describe the general functions of these hormones.
7. Explain how the secretion of each of these hormones is regulated.

Focus Question

How do the functions of the nervous system and the endocrine system differ, and how do they complement one another?

Mastery Test

Now, take the mastery test. Do not guess. As soon as you complete the test, correct it. Note your successes and failures so that you can read the chapter to meet your learning needs.

1. Glands that release their secretions into the blood and help regulate metabolic processes are

   _____ glands.

2. Many hormones are thought to function by acting on receptor sites in the
   a. target cells.
   b. nucleus.
   c. genes.
   d. cytoplasm.

3. Another group of compounds that have hormonelike effects are _____ .

4. The characteristics of the negative feedback system that regulates hormone secretion include
   a. activation by imbalance.
   b. exertion of an inhibitory effect on the gland.
   c. exertion of a stimulating effect on the gland.
   d. a tendency for levels of hormone to fluctuate.

5. The part of the brain most closely related to endocrine function is the _____ .

6. The hormones secreted by the anterior lobe of the pituitary gland include
   a. thyroid-stimulating hormones.
   b. luteinizing hormone.
   c. antidiuretic hormone.
   d. oxytocin.

7. Nerve impulses from the hypothalamus stimulate the _____ lobe of the pituitary gland.

8. The _____ lobe of the pituitary gland is stimulated by releasing factors secreted by the hypothalamus.

9. Which of the following are actions of pituitary growth hormone?
   a. enhance the movement of amino acids through the cell membrane
   b. increase the utilization of glucose by cells
   c. increase the utilization of fats by cells
   d. enhance the movement of potassium across the cell membrane

10. The pituitary hormone that stimulates and maintains milk production following childbirth is

_____ .

11. TSH secretion is regulated by
    a. circulating thyroid hormones.
    b. blood sugar levels.
    c. osmolarity of blood.
    d. TRH secreted by the hypothalamus.

12. Which of the following pituitary hormones help(s) maintain fluid balance?
    a. oxytocin
    b. ACTH
    c. antidiuretic hormone

13. The thyroid hormones that affect the metabolic rate are _____ and

_____ .

14. Which of the following are functions of thyroid hormones?
    a. control sodium levels
    b. decrease rate of energy release from carbohydrates
    c. increase protein synthesis
    d. accelerate growth in children

15. The element necessary for normal function of the thyroid gland is _____ .

16. The thyroid hormone that tends to keep calcium in the bone is _____ .

17. Which of the following statements about parathormone (parathyroid hormone) is (are) true?
    a. Parathormone enhances absorption of phosphorus and calcium from the intestine.
    b. Parathormone stimulates the bone to release ionized calcium.
    c. Parathormone stimulates the kidneys to conserve calcium.
    d. Parathormone secretion is stimulated by the hypothalamus.

18. Injury to or removal of parathyroid glands is likely to result in
    a. reduced osteoclastic activity.
    b. Cushing's disease.
    c. kidney stones.
    d. hypocalcemia.

19. The hormones of the adrenal medulla are _____ and _____ .

20. The adrenal hormone aldosterone belongs to a category of cortical hormones called
    a. mineralocorticoids.
    b. glucocorticoids.
    c. sex hormones.

21. The actions of cortisol include
    a. the breakdown of stored protein to increase the levels of circulating amino acids.
    b. increasing the release of fatty acids and decreasing the use of glucose.
    c. stimulation of gluconeogenesis.
    d. conservation of water.

22. Adrenal sex hormones are primarily (male, female).

23. The endocrine portion of the pancreas are cells called _____ _____

_____ .

24. The hormone that responds to a low blood sugar by stimulating the liver to convert glycogen to glucose is

_____ .

25. The action of insulin that most directly leads to lowered blood sugar levels is
    a. enhancing glucose absorption from the small intestine.
    b. facilitating the transport of glucose across the cell membrane.
    c. promoting the transport of amino acids out of the cell.
    d. increasing the synthesis of fats.

26. The hormone melatonin is secreted by the
    a. thymus.
    b. pineal gland.
    c. gonads.

Study Activities

I. Aids to Understanding Words

Define the following word parts. (p. 277)

-crin

diuret-

endo-

exo-

hyper-

hypo-

para-

toc-

tropic-

II. Introduction, General Characteristics of the Endocrine System, and Hormones and Their Actions (pp. 278–81)

A. Answer these questions concerning the endocrine system. (p. 278)
   1. List the differences between endocrine and exocrine glands.

   2. As a group, endocrine glands regulate _____ _____ .
B. Answer these questions concerning hormones and their actions. (pp. 278–81)
   1. The specific site of a hormone's action is called its _____
   _____ .
   2. Describe how hormones act on their target tissue.

   3. Compare the characteristics of steroid and nonsteroid hormones.

   4. What are prostaglandins?

III. Control of Hormonal Secretions (pp. 281–82)

A. Describe how the negative feedback system regulates hormone secretion. (p. 281)

B. Describe how the positive feedback system regulates hormone secretion. (p. 282)

C. How does the nervous system control hormone secretion? (p. 282)

IV. The Pituitary Gland (pp. 282–85)

A. Describe the structure and location of the pituitary gland. (p. 282)

B. Fill in the following chart. (pp. 283–85)

Hormones of the pituitary gland		
Hormone	Stimulus for Secretion	Actions (Be Specific)
Anterior lobe		
Growth hormone		
Prolactin		
Thyroid-stimulating hormone		
Adrenocorticotropic hormone		
Follicle-stimulating hormone		
Luteinizing hormone		
Posterior lobe		
Antidiuretic hormone		
Oxytocin		

V. The Thyroid Gland (pp. 286–87)

A. On the accompanying illustration, label the thyroid gland, isthmus, larynx, colloid, and follicle. (p. 286)

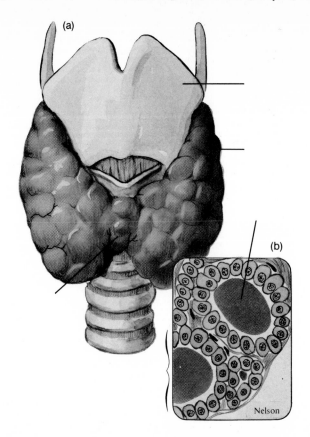

(a)

(b)

Nelson

B. Answer these questions concerning thyroid hormones and their functions. (pp. 286–87)
 1. What element is needed to synthesize thyroxine and triiodothyronine?

 2. What are the functions of thyroxine and triiodothyronine?

 3. What is the function of calcitonin?

VI. The Parathyroid Glands (pp. 288–89)

A. Where are the parathyroid glands located? (p. 288)

B. Describe how parathormone affects blood levels of calcium and phosphorus. Include its effect on bone, the intestine, and the kidneys. (p. 288)

VII. The Adrenal Glands (pp. 289–91)

A. Where are the adrenal glands located? (p. 289)

B. Answer these questions concerning hormones of the adrenal medulla. (pp. 289–90)
   1. List the hormones secreted by the adrenal medulla.

   2. What are the effects of these hormones?

C. Fill in the following chart. (pp. 290–91)

Adrenocortical hormones

	Zone of the Cortex	Stimulus for Secretion	Effects of Hormone
Mineralocorticoids (e.g., aldosterone)			
Glucocorticoids (e.g., cortisol)			
Sex hormones			

D. Describe the negative feedback mechanism that controls the release of cortisol. (p. 291)

VIII. The Pancreas (pp. 292–93)

A. Where is the pancreas located? (p. 292)

B. Fill in the following chart. (p. 292)

Hormones of the pancreas

	Source of Control	Effects of Hormone
Glucagon		
Insulin		

IX. Other Endocrine Glands (pp. 293–94)

A. Where is the pineal gland located, and what is its function? (pp. 293–94)

B. Where is the thymus gland located, and what is its function? (p. 294)

When you have completed the study activities to your satisfaction, retake the mastery test and compare your performance with your initial attempt. If there are still areas you do not understand, repeat the appropriate study activities.

12 Digestion and Nutrition

Overview

This chapter is about the digestive system, which processes food so that nutrients can be absorbed and used by the cells. It names the organs of the digestive system and describes their locations and functions (objectives 1 and 2). The structure, movement, and digestive mechanisms of the alimentary canal are explained (objectives 3 and 4). The function and regulation of the secretions of the digestive organs are introduced (objectives 5 and 6). This chapter also tells how nutrients are absorbed (objective 7). This chapter also presents some basic concepts of nutrition. It describes the use, function, and sources of protein, carbohydrates, lipids, vitamins, and minerals (objectives 8–11). It explains an adequate diet (objective 12).

Study of the digestive system helps in understanding how fuel is made available for metabolism. Knowledge of what foods to select and proper quantities of food provides a firm foundation for the study of nutrition.

Chapter Objectives

After you have studied this chapter, you should be able to

1. Name and describe the location of organs of the digestive system and their major parts.
2. Describe the general functions of each digestive organ and the liver.
3. Describe the structure of the wall of the alimentary canal.
4. Explain how contents of the alimentary canal are mixed and moved.
5. List the enzymes secreted by various digestive organs and describe the function of each enzyme.
6. Describe how digestive secretions are regulated.
7. Explain how products of digestion are absorbed.
8. List the major sources of carbohydrates, lipids, and proteins.
9. Describe how carbohydrates, lipids, and proteins are utilized by cells.
10. List the fat-soluble and water-soluble vitamins and summarize the general functions of each vitamin.
11. List the major minerals and trace minerals and summarize the general functions of each mineral.
12. Describe an adequate diet.

Focus Question

How does the body make the nutrients in a ham sandwich available for use by the cells?

Mastery Test

Now take the mastery test. Do not guess. As soon as you complete the test, correct it. Note your successes and failures so that you can read the chapter to meet your learning needs.

1. The mouth, pharynx, esophagus, stomach, and large and small intestines make up the _____ _____ of the digestive system.

2. The salivary glands, liver, gallbladder, and pancreas are considered _____ _____ .

3. The two basic types of movement of the alimentary canal are _____ movements and _____ movements.

4. Does statement *a* explain statement *b*? _____
   a. Peristalsis is stimulated by stretching the alimentary tube.
   b. Peristalsis acts to move food along the alimentary tube.

5. The tongue is anchored to the floor of the mouth by a fold of membrane called the _____ .

6. During swallowing, muscles draw the soft palate and uvula upward to
   a. move food into the esophagus.
   b. enlarge the area to accommodate a bolus of food.
   c. separate the oral and nasal cavities.
   d. move the uvula from the path of the food bolus.

7. The material that covers the crown of the teeth is
   a. cementum.
   b. dentin.
   c. enamel.
   d. plaque.

8. Which of the following is *not* a function of saliva?
   a. cleansing of mouth and teeth
   b. dissolve chemicals necessary to tasting food
   c. help in formation of food bolus
   d. begin digestion of protein

9. Stimulation of salivary glands by parasympathetic nerves will (increase, decrease) production of saliva.

10. The salivary glands that secrete amylase are the
    a. submaxillary glands.
    b. parotid glands.
    c. sublingual glands.

11. When food enters the esophagus, it is transported to the stomach by a movement called

    _____ .

12. The area of the stomach that acts as a temporary storage area is the
    a. cardiac region.
    b. fundic region.
    c. body region.
    d. pyloric region.

13. The chief cells of the gastric glands secrete
    a. mucus.
    b. hydrochloric acid.
    c. digestive enzymes.
    d. potassium chloride.

14. The digestive enzyme pepsin secreted by gastric glands begins the digestion of
    a. carbohydrate.
    b. protein.
    c. fat.

15. The intrinsic factor secreted by the stomach aids in the absorption of _____ from the small intestine.

16. The release of gastrin is stimulated by
    a. the sympathetic nervous system.
    b. the presence of alkaline substances.
    c. the sight and smell of food.
    d. the presence of such things as protein, caffeine, and alcohol.

17. The presence of food in the small intestine (inhibits, increases) gastric secretion.

18. The semifluid paste formed in the stomach by mixing food and gastric secretions is _____ .

19. The foods that stay in the stomach the longest are high in
    a. fats.
    b. protein.
    c. carbohydrate.

20. Pancreatic enzymes travel along the pancreatic duct and empty into the
    a. duodenum.
    b. jejunum.
    c. ileum.

21. Which of the following enzymes is present in secretions of the mouth, stomach, and pancreas?
    a. amylase
    b. lipase
    c. trypsin
    d. lactase

22. Which of the following is(are) secreted by the pancreas in an inactive form and is activated by a duodenal enzyme?
    a. nuclease
    b. trypsin
    c. chymotrypsin
    d. carboxypeptidase

23. The secretions of the pancreas are (acid, alkaline).

24. The liver is located in the _____ _____ quadrant of the abdomen.

25. Nutrients are brought to the liver cells via the
   a. central vein.
   b. liver capillaries.
   c. hepatic sinusoids.
   d. connective tissue of the lobes of the liver.

26. The only substances in bile that have a digestive function are _____

   _____ .

27. The function(s) of the gallbladder is(are) to
   a. store bile.
   b. secrete bile.
   c. activate bile.
   d. concentrate bile.

28. Which of the following is(are) the function(s) of bile?
   a. emulsification of fat globules
   b. absorption of fats
   c. increase the solubility of amino acids
   d. absorption of fat-soluble vitamins

29. List the portions of the small intestine: _____ , _____ , and

   _____ .

30. The velvety appearance of the lining of the small intestine is due to the presence of
   a. cilia.
   b. villi.
   c. mucus secreted by the small intestine.
   d. capillaries.

31. The intestinal enzyme that breaks down carbohydrates is
   a. sucrase.
   b. maltase.
   c. lipase.
   d. intestinal amylase.

32. The small intestine joins the large intestine at the _____ .

33. The only significant secretion of the large intestine is
   a. potassium.
   b. mucus.
   c. chyme.
   d. water.

34. The only nutrients normally absorbed in the large intestine are _____ and

   _____ .

35. Nutrients, such as amino and fatty acids, that are necessary for health but cannot be synthesized by the body
   are called _____ _____ .

36. Carbohydrates are ingested in such foods as
   a. meat and seafoods.
   b. bread and pasta.
   c. butter and margarine.
   d. milk and cheese.

37. Glucose can be stored as glycogen in the
   a. blood plasma.
   b. muscles.
   c. connective tissue.
   d. liver.

38. The organ most dependent on an uninterrupted supply of glucose is the
   a. heart muscle.
   b. liver.
   c. adrenal gland.
   d. brain.

39. Triglycerides are broken down into _____ _____ _____ and

   _____ .

40. An essential fatty acid that cannot be synthesized by the body is _____

   _____ .

41. Which of the following lipids is found in bile salts?
    a. cholesterol
    b. phospholipids
    c. sterols
    d. vitamin A

42. A lipid that furnishes molecular components for the synthesis of sex hormones and some adrenal hormones is

    _____ .

43. Proteins function as
    a. enzymes that regulate metabolic reactions.
    b. promoters of calcium absorption.
    c. energy supplies.
    d. structural materials in cells.

44. Proteins are absorbed and transported to cells as _____ _____ .

45. A protein that contains adequate amounts of the essential amino acids is called a(n) _____ protein.

46. Does statement *a* explain statement *b*? _____
    a. Carbohydrate is used before other nutrients as an energy source.
    b. Carbohydrate has a protein-sparing effect.

47. Protein requirements are relatively high during
    a. pregnancy.
    b. young adulthood.
    c. early childhood.
    d. old age.

48. Vitamins A, D, E, and K are _____-soluble.

49. Vitamin D_3 is produced by exposure of 7-dehydrocholesterol in the skin to _____ .

50. The vitamin essential for normal blood clotting is
    a. vitamin C.
    b. vitamin K.
    c. vitamin E.
    d. vitamin B.

51. Which of the B-complex vitamins can be synthesized from tryptophan by the body?
    a. niacin
    b. thiamine
    c. riboflavin
    d. pyridoxine

52. Which of the following is *not* a good source of vitamin C?
    a. potatoes
    b. tomatoes
    c. grains
    d. cabbage

53. The most abundant minerals in the body are _____ and _____ .

54. Calcium is necessary for all but which one of the following?
    a. blood coagulation
    b. muscle contraction
    c. color vision
    d. nerve impulse conduction

55. Chlorine is usually ingested with _____ .

56. The best natural source of iron is
    a. liver.
    b. red meat.
    c. egg yolk.
    d. raisins.

57. A substance necessary for bone development, melanin production, and myelin formation is
    a. iron.
    b. iodine.
    c. copper.
    d. zinc.

Study Activities

I. Aids to Understanding Words

Define the following word parts. (p. 303)

aliment-   lingu-

chym-   nutri-

decidu-   peri-

gastr-   pylor-

hepat-   vill-

II. Introduction and General Characteristics of the Alimentary Canal (pp. 304–6)

A. Label the diagram of the digestive system. (p. 304)

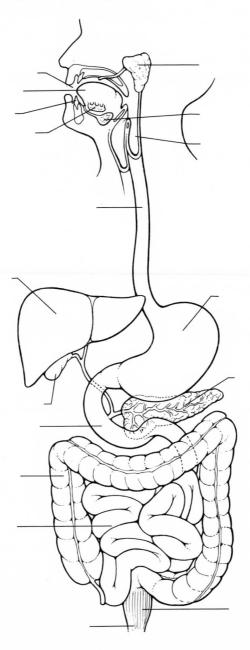

B. Fill in the following chart. (pp. 304–6)

Structure and function of the alimentary tube	
Structure	**Function**
Mucous membrane	
Submucosa	
Muscular layer	
Serous layer	

C. The two types of movement of the alimentary tube are _____ and _____ . (p. 306)

III. Mouth (pp. 306–9)

A. What are the functions of the mouth? (p. 306)

B. What is the function of the tongue? (p. 306)

C. Answer these questions concerning the palate. (pp. 306–7)
   1. What are the parts of the palate?

   2. What is the function of the palate?

D. In the accompanying illustration, label the crown, root, enamel, dentin, gingiva, pulp cavity, cementum, alveolar bone, and root canal. (p. 309)

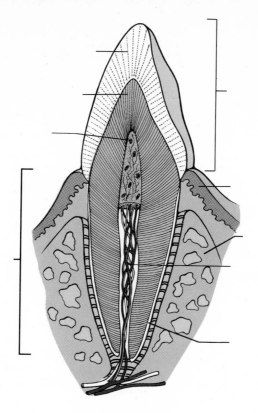

E. Describe the function of the following kinds of teeth. (p. 308)
incisors

cuspids

bicuspids

molars

IV. Salivary Glands (pp. 309–10)

A. Answer these questions concerning salivary glands and their secretions. (p. 309)
   1. What is the function of salivary glands?

   2. What stimulates them to secrete saliva?

B. Fill in the following chart. (pp. 309–10)

Salivary glands

Glands	Location of the Glands	Secretion
Parotid		
Submaxillary		
Sublingual		

V. Pharynx and Esophagus (pp. 310–11)

A. Describe the nasopharynx, oropharynx, and laryngopharynx. In the description, include the muscles of the pharynx. (p. 310)

B. List the events of swallowing. (p. 311)

C. Describe the structure and functions of the esophagus. (p. 311)

VI. Stomach (pp. 311–14)

A. Answer these questions concerning the stomach and its parts. (p. 311)
    1. What are the functions of the stomach?

2. Label the indicated regions of the stomach and identify the function of each region: fundic region, cardiac region, body, pyloric region, pyloric canal, duodenum, pyloric sphincter.

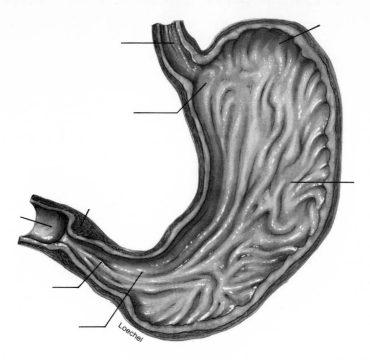

Loechel

B. Fill in the following chart. (p. 312)

Secretions of gastric gland

Cell Type	Secretions	Function and Action
Mucous cell		
Chief cell		
Parietal cell		

C. What substances are absorbed from the stomach? (p. 314)

D. Answer these questions concerning filling and emptying actions. (p. 314)
   1. What is chyme, and how is it produced?

   2. What factors affect the rate at which the stomach empties?

VII. Pancreas (pp. 314–15)

A. Where is the pancreas located? (p. 314)

B. Answer these questions concerning pancreatic juice. (p. 315)
   1. Describe the action of the following pancreatic enzymes.
      amylase

      lipase

      trypsin

      chymotrypsin

      carboxypeptidase

   2. What substance makes the pancreatic juice alkaline?

   3. How does secretin affect pancreatic juice?

   4. How does cholecystokinin affect pancreatic juice?

VIII. Liver and Gallbladder (pp. 315–20)

A. Describe the location of the liver and explain its digestive function. (pp. 315–16)

B. Answer these questions concerning bile. (p. 318)
   1. Describe the composition of bile.

   2. Which of the substances in bile is active in the digestive process?

   3. What is the source of bile pigments?

C. Answer these questions concerning the gallbladder. (pp. 318–20)
   1. Where is the gallbladder located?

   2. How is bile stored?

D. Describe how bile is released. (p. 319)

E. Answer these questions concerning the digestive functions of bile salts. (p. 319)
    1. The digestive function of bile is the _____ of fats.
    2. Bile salts aid the absorption of _____ _____ and
        _____ .

IX. Small Intestine (pp. 320–24)

A. Name and locate the three portions of the small intestine. (pp. 320–21)

B. Label these structures in the accompanying illustration and explain the function of each: lacteal, capillary network, intestinal gland, goblet cells, arteriole, venule, lymph vessel, villus. (p. 321)

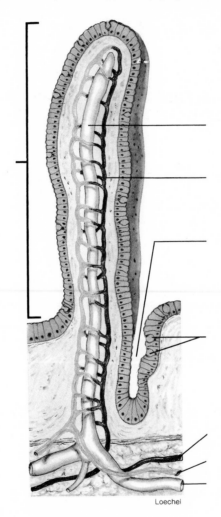

Loechel

C. Answer these questions concerning absorption in the small intestine. (pp. 322–23)
   1. Fill in the following chart.

Digestion and absorption of nutrients in the small intestine

Food	Intestinal Enzyme	Nutrient	Absorption Mechanism	Means of Transport
Carbohydrate				
Protein				
Fat				

D. Answer these questions concerning the movements of the small intestine. (p. 324)
   1. The normal movements of the small intestine are _____ and _____ .

   2. What separates the small and large intestine? How does this structure work?

X. Large Intestine (pp. 325–27)

A. Answer the questions concerning the large intestine and its parts. (p. 325)

    1. Label the parts of the large intestine on the accompanying drawing: ileum, vermiform appendix, cecum, ascending colon, transverse colon, descending colon, ileocecal valve, sigmoid colon, rectum, anal canal.

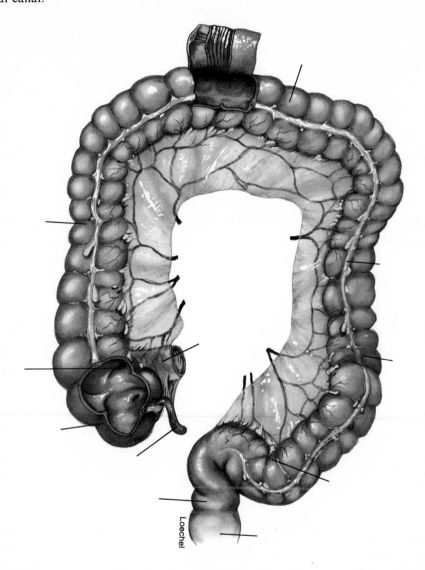

Loechel

    2. How is the structure of the wall of the large intestine different from the structure of the wall of the small intestine?

B. Answer these questions concerning the functions of the large intestine. (p. 326)

    1. What is the function of mucus in the large intestine?

    2. What substances are absorbed in the large intestine?

C.   Answer these questions concerning the movements of the large intestine. (pp. 326–27)
  1.   Describe the movements of the large intestine.

  2.   List the events of the defecation reflex.

D.   Describe the composition of feces. (p. 327)

XI. Carbohydrates (pp. 327–28)

A.   Nutrients needed for health that cannot be synthesized by the body are called _____
  _____ . (p. 327)

B.   Answer these questions concerning the sources and the use of carbohydrates. (p. 327)
  1.   Carbohydrates are _____ compounds that are used primarily to supply
  _____ .

  2.   In what forms are carbohydrates ingested?

  3.   In what forms are carbohydrates absorbed?

  4.   What form of carbohydrates is most commonly used by the cell as fuel?

  5.   Identify the areas in which glucose is stored for rapid mobilization.

C.   Answer these questions concerning carbohydrate requirements. (p. 328)
  1.   What is the function of carbohydrate?

  2.   What cells are particularly dependent on a continuous supply of glucose?

XII. Lipids (pp. 328–29)

A.   In what forms are lipids usually ingested? (p. 328)

B.   Answer these questions concerning the use of lipids. (pp. 328–29)
  1.   Describe the role of the liver in the use of lipids.

2. An essential fatty acid that cannot be synthesized by the body is _____
_____ .

3. Describe the role of adipose tissue in the use of lipids.

4. What is the function of lipids?

XIII. Proteins (pp. 329–30)

A. Answer these questions concerning sources of protein. (pp. 329–30)
   1. In what form is protein transported and used by cells?

   2. What is an essential amino acid?

   3. What is the difference between a complete and an incomplete protein?

B. Answer these questions concerning the use of amino acids. (p. 330)
   1. Describe the various ways body cells may use amino acids.

   2. How are proteins used by the body?

XIV. Vitamins (p. 331)

A. Answer these questions concerning vitamins. (p. 331)
   1. What is a vitamin?

   2. What are the general characteristics of fat-soluble vitamins?

   3. What are the general characteristics of water-soluble vitamins?

B. Fill in the following chart. (p. 331)

Vitamins			
Vitamin	**Characteristics**	**Functions**	**Sources**
Vitamin A			
Vitamin D			
Vitamin E			
Vitamin K			
Vitamin B Complex Thiamine (B₁)			
Riboflavin (B₂)			
Niacin (nicotinic acid)			
Pyridoxine (B₆)			
Pantothenic acid			
Cyanocobalamin (B₁₂)			
Folacin (folic acid)			
Biotin			
Ascorbic acid (vitamin C)			

XV. Minerals (pp. 332–34)

  A. Describe the general characteristics of minerals. (pp. 332–33)

B. Fill in the following chart. (p. 333)

Major minerals

Mineral	Distribution	Regulatory Mechanism	Function	Source
Calcium				
Phosphorus				
Potassium				
Sulfur				
Sodium				
Chlorine				
Magnesium				

C. Answer these questions concerning trace elements. (p. 334)
  1. What is a trace element?

  2. List the six trace elements and identify the function of each.

XVI. Adequate Diets (pp. 334–36)

A. What is an adequate diet? (p. 334)

B. Why is it impossible to design a diet that is adequate for everyone? (p. 334)

When you have completed the study activities to your satisfaction, retake the mastery test and compare your performance with your initial attempt. If there are still areas you do not understand, repeat the appropriate study activities.

13 Respiratory System

Overview

The respiratory system permits the exchange of oxygen, which is needed for cellular metabolism, and carbon dioxide, a by-product of cellular metabolism. This chapter describes the location and function of the organs of the respiratory system and discusses how they contribute to the overall function of the system (objectives 1–3). Respiratory air movements, respiratory volumes, and how normal breathing is controlled are discussed (objectives 4–7). Gas exchange and transport are explained (objectives 8 and 9).

Air must be taken into the lungs so that oxygen and carbon dioxide can be exchanged. An understanding of these events and how they are controlled is basic to understanding how cells produce energy for life processes.

Chapter Objectives

After you have studied this chapter, you should be able to

1. List the general functions of the respiratory system.
2. Name and describe the location of the organs of the respiratory system.
3. Describe the functions of each organ of the respiratory system.
4. Explain how inspiration and expiration are accomplished.
5. Name and define each of the respiratory air volumes.
6. Locate the respiratory center and explain how it controls normal breathing.
7. Discuss how various factors affect the respiratory center.
8. Describe the structure and function of the respiratory membrane.
9. Explain how oxygen and carbon dioxide are transported in blood.

Focus Question

How do the lungs extract oxygen from a mixture of gases in the atmosphere and give off waste gases from the body?

Mastery Test

Now take the mastery test. Do not guess. As soon as you complete the test, correct it. Note your successes and failures so that you can read the chapter to meet your learning needs.

1. Match the functions in the first column with the appropriate part of the nose in the second column.

    ____ a. warm incoming air
    ____ b. trap particulate matter in the air
    ____ c. prevent infection
    ____ d. moisten air
    ____ e. move nasal secretions to pharynx

    1. mucous membrane
    2. mucus
    3. cilia

2. The pharynx is the cavity behind the mouth extending from the _____
    _____ to the _____ .

3. The portions of the larynx concerned with preventing foreign objects from entering the trachea are the
    a. arytenoid cartilages.
    b. glottis.
    c. epiglottis.
    d. hyoid bone.

4. The trachea is maintained in an open position by
    a. cartilaginous rings.
    b. the amount of collagen in the wall.
    c. the tone of smooth muscle in the wall of the trachea.
    d. the continuous flow of air through the trachea.

5. The right and left bronchi arise from the trachea at
   a. the suprasternal notch.
   b. the manubrium of the sternum.
   c. the fifth thoracic vertebra.
   d. the eighth intercostal space.

6. The smallest branches of the bronchial tree are the _____ _____ .

7. The serous membrane covering the lungs is the _____ _____ .

8. The serous membrane covering the inner wall of the thoracic cavity is the _____

   _____ .

9. Inspiration occurs after the diaphragm _____ , thus (increasing, decreasing) the size of

   the thorax and (increasing, decreasing) the pressure within the thorax.

10. The other muscles that normally act to change the size of the thorax are
    a. the sternocleidomastoids.
    b. the pectorals.
    c. the intercostals.
    d. the latissimus dorsi.

11. Expansion of the lungs during inspiration is assisted by the surface tension of fluid in the

    _____ cavity.

12. The surface tension of fluid in the alveoli is decreased by a secretion, _____ , which

    prevents collapse of the alveoli.

13. The force responsible for expiration comes mainly from
    a. contraction of intercostal muscles.
    b. change in the surface tension within alveoli.
    c. elastic recoil of tissues in the lung and thoracic wall.
    d. contraction of abdominal muscle to push the
       diaphragm upward.

14. The pressure in the thoracic cavity during inspiration is
    a. greater than atmospheric pressure.
    b. less than atmospheric pressure.
    c. the same as atmospheric pressure.

15. The amount of air that enters and leaves the lungs during a normal, quiet respiration is the
    a. vital capacity.
    b. inspiratory reserve volume.
    c. total lung capacity.
    d. tidal volume.

16. Normal breathing is controlled by the respiratory center located in the _____

    _____ .

17. The inflation reflexes are activated by
    a. stretch receptors in the bronchioles and alveoli.
    b. an increase in hydrogen ions.
    c. a decrease in oxygen saturation.
    d. a sudden fall in blood pressure.

18. The strongest stimulus to increase respiratory rate and depth is to increase the blood concentration of

    _____ _____ .

19. The respiratory membrane consists of a single layer of epithelial cells and basement membrane between a(n)

    _____ and a(n) _____ .

20. The rate at which a gas diffuses from one area to another is determined by differences in

    _____ in the two areas.

21. The pressure of each gas within a mixture is known as its _____

    _____ .

22. Oxygen is transported to cells by combining with _____ .

23. The largest amount of carbon dioxide is transported
    a. dissolved in blood.
    b. combined with hemoglobin.
    c. as bicarbonate.
    d. as carbonic anhydrase.

Study Activities

I. Aids to Understanding Words

Define the following word parts. (p. 343)

alveol-   cric-

bronch-   epi-

                                           hem-

II. Organs of the Respiratory System (pp. 344–50)

A. In the accompanying drawing, label these structures: nasal cavity, nostril, mouth cavity, pharynx, larynx, trachea, bronchus, right lung, left lung, epiglottis, hard palate, soft palate. (p. 344)

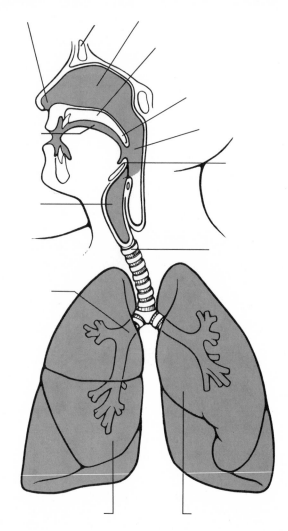

B. Match the functions in the first column with the appropriate terms in the second column. (pp. 344–45)

____ 1. entrap dust   a. mucous membrane
____ 2. lighten skull and provide vocal resonance   b. mucus
____ 3. warm and humidify air entering the nose   c. sinuses
____ 4. provide movement to mucus layer   d. cilia

C. What is the location and function of the pharynx? (p. 345)

D. Answer these questions concerning the larynx. (p. 345–46)
    1. Identify these structures in the accompanying drawings: hyoid bone, epiglottic cartilage, trachea, thyroid cartilage, cricoid cartilage.

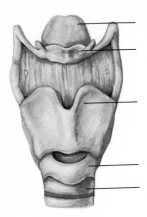
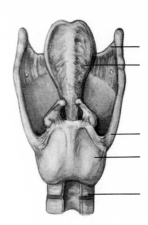

    2. The structure of the larynx that helps close the glottis during swallowing is the

        _____ .

    3. The structures of the larynx that produce sound are the _____ .

E. The trachea is prevented from collapsing by the presence of _____
    _____ that are C-shaped. (p. 347)

F. Answer these questions concerning the bronchial tree. (pp. 347–48)
    1. On the accompanying drawing, label the larynx, trachea, right primary bronchus, secondary bronchus, tertiary bronchi, alveolary duct, aveoli.

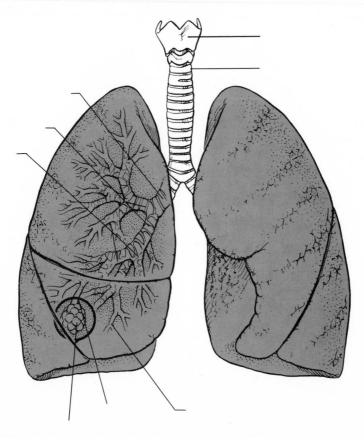

2.   What is the function of alveoli?

G.   Answer these questions concerning the lungs. (pp. 348–49)
     1.   The right lung has _____ lobes and the left lung has
          _____lobes.
     2.   How is the pleural cavity formed?

III. Mechanism of Breathing (pp. 350–53)

A.   Answer these questions concerning inspiration. (pp. 350–51)
     1.   List the events of inspiration beginning with stimulation of the phrenic nerves.

     2.   Describe the role of surface tension in the pleural cavity and in the alveoli.

B.   List the events of expiration. (p. 352)

C.   Match the terms in the first column with the correct definition in the second column. (pp. 352–53)
     ____ 1. inspiratory reserve volume   a. volume moved in or out of the lungs during quiet
     ____ 2. expiratory reserve volume   respiration
     ____ 3. residual volume   b. volume that can be inhaled during forced breathing
     ____ 4. vital capacity   in addition to tidal volume
     ____ 5. total lung capacity   c. volume that can be exhaled in addition to tidal
     ____ 6. tidal volume   volume
                                                 d. volume that remains in the lungs at all times
                                                 e. maximum air that can be exhaled after taking the
                                                    deepest possible breath
                                                 f. total volume of air the lungs can hold

IV. Control of Breathing (pp. 354–56)

A. Answer these questions concerning the respiratory center. (pp. 354–55)

   1. Label the indicated structures: vagus nerve, sensory pathway, respiratory center, motor pathways, spinal cord, rib, phrenic nerve, alveoli, intercostal nerve, external intercostal muscles, diaphragm, lung.

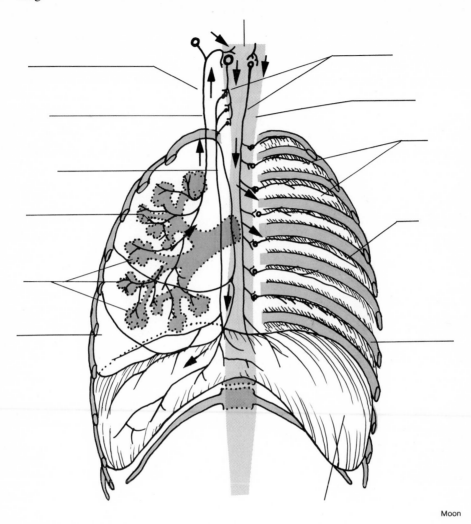

Moon

   2. Describe the function of the respiratory center in maintaining normal breathing. Include the medullary rhythmicity area, dorsal respiratory group, ventral respiratory group, and pneumotaxic area.

B. How do chemical factors affect breathing? (p. 355)

V. Alveolar Gas Exchanges (pp. 357–58)

A. Identify the layers of cells that make up the respiratory membrane. (p. 357)

B. Answer these questions concerning diffusion through the respiratory membrane. (pp. 357–58)
   1. What determines the direction and rate at which gases diffuse from one area to another?

   2. Define *partial pressure of gas.*

   3. Use partial pressure of gas to explain the exchange of oxygen and carbon dioxide in the alveoli.

VI. Transport of Gases   (pp. 358–60)

A. Answer these questions concerning oxygen transport. (pp. 358–59)
   1. Describe how oxygen is transported to cells.

   2. Why is oxygen released at the cell?

B. A gas that interferes with oxygen transport by forming a stable bond with hemoglobin is
   _____ _____ . (p. 359)

C. Answer these questions concerning carbon dioxide transport. (pp. 359–60)
   1. How is carbon dioxide transported away from the cell?

   2. Why is CO_2 lost in the lung?

When you have completed the study activities to your satisfaction, retake the mastery test and compare your performance with your initial attempt. If there are still areas you do not understand, repeat the appropriate study activities.

14 Blood

Overview

This chapter deals with a major connective tissue—the blood. It describes the general characteristics and major functions of blood (objective 1). It identifies the various types of blood cells and the components and functions of plasma (objectives 2 and 4). It explains blood typing, control of red blood cell production, blood coagulation, and reaction to mixing of blood types (objectives 3, 5–8).

Knowledge of the circulatory system expands the concept of how cells meet their need for oxygen. This knowledge also helps in understanding how the body recognizes and rejects foreign protein.

Chapter Objectives

After you have studied this chapter, you should be able to

1. Describe the general characteristics of blood and discuss its major functions.
2. Distinguish between the various types of cells found in blood.
3. Explain how red blood cell production is controlled.
4. List the major components of blood plasma and describe the functions of each.
5. Define *hemostasis* and explain the mechanisms that help to achieve it.
6. Review the major steps in blood coagulation.
7. Explain the basis for blood typing and how it is used to avoid adverse reactions following blood transfusions.
8. Describe how blood reactions may occur between fetal and maternal tissues.

Focus Question

How does the structure of the blood help meet oxygenation needs, allow recognition and rejection of foreign protein, and control coagulation of the blood?

Mastery Test

Now take the mastery test. Do not guess. As soon as you complete the test, correct it. Note your successes and failures so that you can read the chapter to meet your learning needs.

1. The intercellular material of blood is _____ .

2. Plasma represents _____% of a normal blood sample.

3. Damaged red blood cells are destroyed by reticuloendothelial cells called
   a. leukocytes.
   b. macrophages.
   c. neutrophils.
   d. granulocytes.

4. The heme portion of damaged red blood cells is decomposed into iron and
   a. biliverdin.
   b. bilirubin.
   c. bile.

5. Red blood cells are produced in
   a. the spleen.
   b. red marrow.
   c. yellow marrow.
   d. the liver.

6. Red blood cell production is stimulated by a hormone, _____ , that is released from the kidney in response to low oxygen concentration.

7. Does statement *a* explain statement *b*? _____
   a. Vitamin B$_{12}$ and folic acid are necessary to cell growth and reproduction.
   b. The rate of red blood cell reproduction makes this process especially dependent on vitamin B$_{12}$ and folic acid.

8. An inadequate supply of iron in the diet may lead to a type of anemia called
   a. hypochromic anemia.
   b. aplastic anemia.
   c. hemorrhagic anemia.
   d. hemolytic anemia.

9. The most numerous type of white blood cell is the
   a. neutrophil.
   b. eosinophil.
   c. monocyte.
   d. lymphocyte.

10. The white blood cell that forms antibodies necessary for immunity to specific diseases is the
    a. basophil.
    b. lymphocyte.
    c. thrombocyte.
    d. eosinophil.

11. The normal white blood cell count is _____ to _____ per cubic millimeter (mm^3) of blood.

12. The blood element concerned with control of bleeding and formation of clots is the _____ .

13. Match the functions and characteristics in the first column with the appropriate plasma proteins from the second column.
    ____ a. largest molecular size
    ____ b. significant in maintaining osmotic pressure
    ____ c. transports lipids and fat-soluble vitamins
    ____ d. antibody(ies) of immunity
    ____ e. plays a part in blood clotting

    1. albumins
    2. globulins
    3. fibrinogen

14. The most abundant plasma electrolytes are
    a. calcium.
    b. sodium.
    c. potassium.
    d. chlorides.

15. A platelet plug begins to form when platelets are
    a. exposed to air.
    b. exposed to a rough surface.
    c. exposed to calcium.
    d. crushed.

16. The basic event in the formation of a blood clot is the transformation of a soluble plasma protein, _____ , to a relatively insoluble protein, _____ .

17. Substances believed necessary to activate prothrombin are thought to include
    a. calcium ions.
    b. potassium ions.
    c. phospholipids.
    d. glucose.

18. Prothrombin is a plasma protein that is produced by
    a. the kidney.
    b. the small intestine.
    c. the pancreas.
    d. the liver.

19. Once a blood clot begins to form, it promotes still more clotting. This is an example of a _____ _____ system.

20. A fragment of a blood clot that is traveling in the bloodstream is called a(n) _____ .

21. The clumping together of red blood cells when unlike types of blood are mixed is due to antibodies in the plasma and antigens in the
    a. thrombocytes.
    b. erythrocytes.
    c. basophils.
    d. eosinophils.

22. The person with type A blood has
    a. agglutinogen A and agglutinin B.
    b. agglutinogens A and B.
    c. agglutinins A and B.
    d. neither agglutinin A nor B.

23. Agglutinins for Rh appear
    a. spontaneously as an inherited trait.
    b. only rarely for poorly understood reasons.
    c. only in response to stimulation by Rh agglutinogens.

Study Activities

I. Aids to Understanding Words

Define the following word parts. (p. 367)

agglutin-   leuko-

bil-   -osis

embol-   -poiet

erythr-   -stas

hemo-   thromb-

II. Blood and Blood Cells (pp. 368–74)

A.   Answer these questions concerning the volume and composition of blood. (p. 368)
   1.   List the solids of the blood.

   2.   What is the blood volume of an average (70 kg) male?

   3.   What part of blood tissue is plasma?

B.   Answer these questions concerning red blood cells. (p. 368)
   1.   The shape of a red blood cell is a(n) _____ _____ .
   2.   How does the shape enhance the function of red blood cells?

   3.   Red blood cells are _____ red when carrying oxygen and are

        _____ red when oxygen is released.
   4.   Why does the red blood cell lack a nucleus?

C. Answer these questions concerning red blood cell counts. (p. 369)
    1. What is the normal red blood cell count for a man? A woman?

    2. What factors provoke a normal increase in red blood cells?

D. Answer these questions concerning destruction of red blood cells. (p. 369)
    1. How are red blood cells damaged?

    2. Damaged red blood cells are destroyed by cells called _____ , located in the

        _____ and _____ .

E. Answer these questions concerning red blood cell production and its control. (pp. 369–71)
    1. Where are red blood cells produced?

    2. How is the production of red blood cells controlled?

F. Answer these questions concerning dietary factors affecting red blood cell production. (p. 371–72)
    1. What is the role of folic acid and vitamin B_{12} in red blood cell production?

    2. What is the role of iron in red blood cell production?

G. Fill in the following chart. (pp. 372–73)

Types of white blood cells			
White Blood Cell	**Description**	**Percent of Total**	**Function**
Granulocytes			
Neutrophil			
Eosinophil			
Basophil			
Agranulocytes			
Monocyte			
Lymphocyte			

H. Answer these questions concerning white blood cell counts. (pp. 373–74)
   1. What is a normal white cell count?
   2. What causes an increase or a decrease in white blood cells?

   3. What is a differential white blood cell count?

I. Answer these questions concerning functions of white blood cells. (p. 374)
   1. How do white blood cells protect the body against microorganisms?

2. Describe the inflammatory response. Include the role of histamine and the phenomenon of chemotaxis.

J. Describe the structure and function of platelets. (p. 374)

III. Blood Plasma (pp. 374–77)

A. Fill in the following chart. (pp. 375–77)

Plasma proteins

Protein	Description	Percentage of Total	Function
Albumins			
Globulins			
Fibrinogen			

B. Answer these questions concerning nutrients and gases. (p. 377)
   1. What nutrients are found in plasma?

   2. What gases are found in plasma?

   3. Describe the nonprotein nitrogenous substances found in plasma.

C. Answer these questions concerning plasma electrolytes. (p. 377)
   1. What electrolytes are found in plasma?

   2. What is the function of electrolytes?

IV. Hemostasis (pp. 377–80)

    A.   List the mechanisms of hemostasis. (pp. 377–78)

    B.   How is a platelet plug formed? (p. 378)

    C.   Describe the major events in the blood-clotting mechanism. (pp. 378–79)

    D.   Describe the mechanisms that prevent coagulation. (p. 378)

    E.   What is a thrombus? What is an embolus? What conditions predispose to formation of thrombi? (p. 379)

V. Blood Groups and Transfusions (pp. 380–84)

    A.   Answer these questions concerning agglutinogens and agglutinins. (p. 380)

        1.   What are agglutinogens and agglutinins?

        2.   Agglutinogens are present in the _____ _____

                        _____ ; agglutinins in the _____ .

    B.   Answer these questions concerning the ABO blood group. (pp. 380–82)

        1.   Describe the basis for ABO blood types.

        2.   Why is it unsafe to mix different blood types?

C. Answer these questions concerning the Rh blood group. (pp. 383–84)
   1. What is the Rh factor?

   2. How does this differ from A, B, and O agglutinogens?

   3. Describe a transfusion reaction.

When you have completed the study activities to your satisfaction, retake the mastery test and compare your performance with your initial attempt. If there are still areas you do not understand, repeat the appropriate study activities.

15 Cardiovascular System

Overview

This chapter deals with the system that transports blood to and from cells—the cardiovascular system. It identifies the major organs of the cardiovascular system and explains their functions (objective 1). It discusses the location, structure, and function of the parts of the heart and the major types of blood vessels (objectives 2 and 6). It explains the pulmonary, systemic, and coronary circuits of the cardiovascular system and the major vessels in each circuit (objectives 3, 9, and 10). It discusses the cardiac cycle and its control and relates these to the normal ECG (objectives 4 and 5). In addition, it describes how blood pressure is created and controlled and the mechanism that aids in the return of venous blood to the heart (objectives 7 and 8).

After learning about the respiratory system and the blood, study of the cardiovascular system completes the knowledge of how oxygen is transported to cells and how waste products are transported away from cells.

Chapter Objectives

After you have studied this chapter, you should be able to

1. Name the organs of the cardiovascular system and discuss their functions.
2. Name and describe the location of the major parts of the heart and discuss the function of each part.
3. Trace the pathway of blood through the heart and vessels of the coronary circulation.
4. Discuss the cardiac cycle and explain how it is controlled.
5. Identify the parts of a normal ECG pattern and discuss the significance of this pattern.
6. Compare the structures and functions of the major types of blood vessels.
7. Describe the mechanism that aids in the return of venous blood to the heart.
8. Explain how blood pressure is created and controlled.
9. Compare the pulmonary and systemic circuits of the cardiovascular system.
10. Identify and locate the major arteries and veins of the pulmonary and systemic circuits.

Focus Question

What are the unique characteristics of the cardiovascular system that permit it to transport fluid and gases for the body?

Mastery Test

Now take the mastery test. Do not guess. As soon as you complete the test, correct it. Note your successes and failures so that you can read the chapter to meet your learning needs.

1. The heart is a cone-shaped, muscular pump located within the _____ .

2. The visceral pericardium is also known as the
   a. epicardium.
   b. myocardium.
   c. endocardium.

3. Purkinje fibers are located in the
   a. epicardium.
   b. myocardium.
   c. endocardium.
   d. parietal pericardium.

4. The upper chambers of the heart are the right and left _____ ; the lower chambers are the right and left _____ .

5. The vessels that empty into the upper right chamber of the heart are
   a. inferior and superior venae cavae.
   b. pulmonary veins.
   c. pulmonary artery.
   d. coronary sinus.

6. The valve between the chambers of the left side of the heart is the
   a. semilunar valve.
   c. tricuspid valve.
   b. bicuspid valve (mitral valve).

7. Blood is supplied to the heart by the right and left _____ _____ .

8. Atrial contraction, while the ventricles relax, followed by ventricular contraction, while the atria relax, is known as the _____ _____ .

9. A mass of merging cells that function as a unit is called
   a. smooth muscle.
   c. the sinoatrial node.
   b. functional syncytium.
   d. the cardiac conduction system.

10. The cells that initiate the stimulus for contraction of the heart muscle are located in the
    a. sinoatrial node.
    c. Purkinje fibers.
    b. atrioventricular node.
    d. bundle of His.

11. A recording of the electrical changes that occur in the myocardium during the cardiac cycle is a(n) _____ .

12. In the recording described in question 11, atrial contraction is represented by the
    a. P wave.
    c. T wave.
    b. QRS complex.
    d. U wave.

13. The effect of parasympathetic nerve stimulation on the heart is to (decrease, increase) the heart rate.

14. Abnormalities in the concentration of which of the following ions is likely to interfere with contraction of the heart?
    a. chloride
    c. calcium
    b. potassium
    d. sodium

15. The vessel that participates directly in the exchange of substances between the cell and the blood is the
    a. arteriole.
    c. capillary.
    b. artery.
    d. venule.

16. When the smooth muscle of the artery contracts, the action is called _____ .

17. The amount of blood that flows into capillaries is regulated by
    a. constriction and dilation of capillaries.
    c. the amount of intercellular tissue.
    b. arterioles.
    d. precapillary sphincters.

18. The transport mechanisms used by the capillaries are _____ ,
    _____ , and _____ .

19. Blood pressure is highest in
    a. an artery.
    c. a capillary.
    b. an arteriole.
    d. a vein.

20. Plasma proteins help retain water in the blood by maintaining
    a. osmotic pressure.
    c. a vacuum.
    b. hydrostatic pressure.

21. The middle layer of the walls of veins differs from that of the arteries in that
    a. it contains more connective tissue.
    c. this layer is thicker in the vein.
    b. it contains less smooth muscle.
    d. it contains some striated muscle.

22. Blood in veins is kept flowing in one direction by the presence of _____ .

23. Which of the following vessels carries deoxygenated blood?
    a. aorta
    c. basilar artery
    b. innominate artery
    d. pulmonary artery

24. The pulmonary veins enter the _____ _____ .

25. The aortic bodies containing pressoreceptors and chemoreceptors are located in the
    a. ascending aorta.   c. aortic sinuses.
    b. aortic arch.   d. descending aorta.

26. The veins that drain the abdominal viscera empty into a unique venous system called the

    _____ _____ _____ .

27. The maximum pressure in the artery, occurring during ventricular contraction, is
    a. diastolic pressure.   c. mean arterial pressure.
    b. systolic pressure.   d. pulse pressure.

28. The amount of blood pushed out of the ventricle with each contraction is called _____

    _____ .

29. List the factors that influence blood pressure.

30. Starling's law is related to which of the following cardiac structures?
    a. interventricular septum   c. muscle fibers
    b. conduction system   d. heart valves

31. When the pressoreceptors in the aorta and carotid artery sense an increase in blood pressure, the medulla will
    relay (sympathetic, parasympathetic) impulses.

32. Peripheral resistance is maintained by increasing or decreasing the size of
    a. capillaries.   c. venules.
    b. arterioles.

33. Venous blood flow is maintained by all but which of the following factors?
    a. blood pressure   c. vasoconstriction of veins
    b. skeletal muscle contraction   d. respiratory movements

Study Activities

I. Aids to Understanding Words

Define the following word parts. (p. 391)

brady-   syn-

diastol-   systol-

-gram   tachy-

papill-

II. Introduction, Structure of the Heart, and Actions of the Heart (pp. 392–404)

A. Answer these questions concerning the cardiovascular system and the location of the heart. (p. 392)

1. What is the function of the cardiovascular system?

2. Describe the precise location of the heart.

B. Answer these questions concerning the coverings of the heart. (p. 392)
    1. The heart is enclosed by a double-layered _____ .
    2. What is the function of the fluid in the pericardial space?

C. Answer these questions concerning the wall of the heart. (pp. 392–93)
    1. Label the following drawing: epicardium, myocardium, endocardium, wall of the heart, heart chambers.

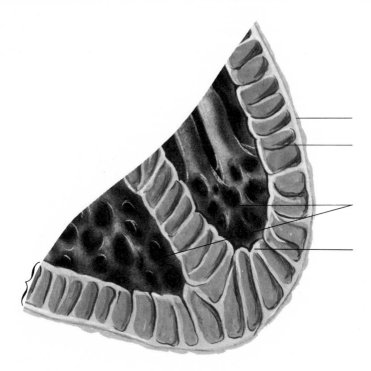

    2. Describe the function of each of the labeled portions of the wall of the heart.

D. Answer these questions concerning heart chambers and valves. (pp. 393–95)
    1. List the chambers of the heart.

2. Label these structures in the accompanying drawing: right and left ventricles, right and left atria, superior and inferior venae cavae, aorta, tricuspid valve, pulmonary semilunar valve, aortic semilunar valve, bicuspid valve, pulmonary veins, pulmonary artery, chordae tendineae, papillary muscles, interventricular septum.

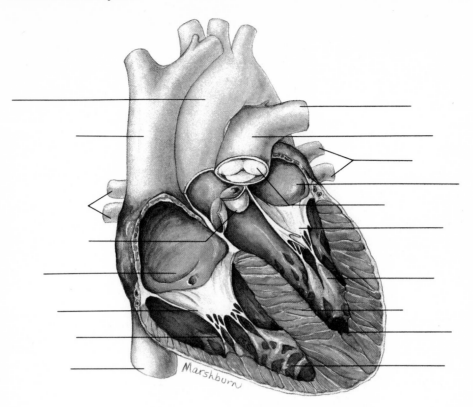

3. What vessels bring blood to the right atrium?

4. What vessels bring blood to the left atrium?

E. Trace the path of the blood through the heart. Include all valves. (pp. 396–98)

F. The cells of the heart are supplied with blood via the _____ _____ . (p. 396)

G. Answer these questions concerning the cardiac cycle. (p. 398)
   1. What events make up a cardiac cycle?

   2. What produces the heart sounds heard with a stethoscope?

H. Describe the characteristics of cardiac muscle fibers. (p. 399)

I. Answer these questions concerning the cardiac conduction system. (pp. 400–401)
    1. Label the parts of the cardiac conduction system in the accompanying drawing: interatrial septum, S-A node, A-V node, A-V bundle, Purkinje fibers, interventricular septum.

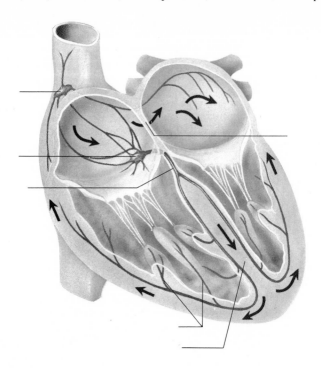

    2. Trace an impulse through the cardiac conduction system.

J. Answer these questions concerning the electrocardiogram. (pp. 401–2)
    1. A recording of electrical changes in the myocardium is a(n) _____ .
    2. What events in the cardiac cycle are represented by each of the following: P wave, QRS complex, T wave?

K. Answer these questions concerning regulation of the cardiac cycle. (pp. 402–3)
    1. How is the heart regulated by autonomic reflexes?

    2. How is the heart affected by potassium and calcium?

III. Blood Vessels (pp. 404–8)

    A.   Answer these questions concerning the cardiovascular system and arteries. (pp. 404–5)

        1.   Describe the closed circuit formed by blood vessels.

        2.   What is the structure of the wall of arteries?

        3.   How is the structure of arterioles different from that of arteries?

    B.   Answer these questions concerning capillaries. (pp. 405–6)

        1.   What is the structure of capillaries?

        2.   What determines the density of capillaries within tissue?

        3.   How is the distribution of blood in the various capillary pathways regulated?

        4.   Describe the following transport mechanisms in the capillary.

        filtration

        osmosis

        diffusion

    C.   Answer these questions concerning veins. (pp. 407–8)

        1.   What is the structure of a vein?

        2.   In what ways is this different from the structure of an artery?

        3.   How do veins function as a blood reservoir?

Blood Pressure (pp. 408–10)

    A.   Answer these questions concerning arterial blood pressure. (p. 408)

        1. What is blood pressure?

        2. What cardiac events are related to systolic and diastolic arterial pressure?

        3. What is the pulse?

    B.   How do each of the following factors influence arterial blood pressure? (pp. 408–9)

       heart action

       blood volume

       peripheral resistance

       viscosity

    C.   Answer these questions concerning control of blood pressure. (pp. 409–10)

        1. How is cardiac output calculated?

        2. Discuss the mechanical, neural, and chemical factors that affect cardiac output.

        3. How is peripheral resistance regulated?

    D.   How do factors such as skeletal muscle contraction, breathing, movements, and vasoconstriction of veins influence venous blood flow? (p. 410)

When you have completed the study activities to your satisfaction, retake the mastery test and compare your performance with your initial attempt. If there are still areas you do not understand, repeat the appropriate study activities.

16 Lymphatic System

Overview

The lymphatic system has two major functions: it helps maintain fluid balance in the tissues of the body and has a major role in the defense against infection. This chapter describes the general functions of the lymphatic system (objective 1). In the discussion of fluid balance, it describes the major lymphatic pathways and lymph formation and circulation (objectives 2–4). In discussing the defense function, it explains lymph nodes and their functions (objective 5). Specific and nonspecific body defenses and the functions of the thymus, spleen, lymphocytes, and immunoglobulins (objectives 6–9) are discussed. Various types of immune responses—primary and secondary responses, active and passive responses, allergic reactions, and tissue rejection reactions (objectives 10–12)—are also explained.

Study of the lymphatic system completes the knowledge of how fluid is transported to and away from tissues. Knowledge of the immune mechanisms of the lymphatic system is the basis for understanding how the body defends itself against specific kinds of threat.

Chapter Objectives

After you have studied this chapter, you should be able to

1. Describe the general functions of the lymphatic system.
2. Describe the location of the major lymphatic pathways.
3. Describe how tissue fluid and lymph are formed and explain the function of lymph.
4. Explain how lymphatic circulation is maintained.
5. Describe a lymph node and its major functions.
6. Discuss the functions of the thymus and spleen.
7. Distinguish between specific and nonspecific body defenses and provide examples of each defense.
8. Explain how two major types of lymphocytes are formed and how they function in immune mechanisms.
9. Name the major types of immunoglobulins and discuss their origins and actions.
10. Distinguish between primary and secondary immune responses.
11. Distinguish between active and passive immunity.
12. Explain how allergic reactions and tissue rejection reactions are related to immune mechanisms.

Focus Question

How is the circulatory function of the lymphatic system related to its role in immune response?

Mastery Test

Now take the mastery test. Do not guess. As soon as you complete the test, correct it. Note your successes and failures so that you can read the chapter to meet your learning needs.

1. The smallest vessels in the lymphatic system are called _____

   _____ .

   The largest vessels are called _____ _____ .

2. The largest lymph vessel is the
   a. lumbar trunk.
   b. thoracic duct.
   c. lymphatic duct.
   d. intestinal trunk.

3. Lymph rejoins the blood and becomes part of the plasma in
   a. lymph nodes.
   b. the right and left subclavian veins.
   c. the inferior and superior venae cavae.
   d. the right atrium.

4. Tissue fluid originates from
   a. the cytoplasm of cells.
   b. lymph fluid.
   c. blood plasma.

5. The function(s) of lymph is (are) to
   a. recapture protein molecules lost in the capillary bed.
   b. form tissue fluid.
   c. transport foreign particles to lymph nodes.
   d. recapture electrolytes.

6. The mechanisms that move lymph through lymph vessels are similar to those that move blood through (arteries, veins).

7. Lymph nodes are shaped like
   a. almonds.
   b. peas.
   c. kidney beans.
   d. convex disks.

8. Compartments within the node contain dense masses of
   a. epithelial tissue.
   b. cilia.
   c. oocytes.
   d. lymphocytes.

9. Which of the following types of cell is produced by lymph nodes?
   a. leukocytes
   b. lymphocytes
   c. eosinophils
   d. basophils

10. The thymus is located in
    a. the posterior neck.
    b. the thorax.
    c. the upper abdomen.
    d. the left pelvis.

11. The thymus produces a substance called _____ , which seems to stimulate the development of _____ tissue.

12. The largest of the lymphatic organs is the _____ .

13. Which of the following statements about the spleen is (are) true?
    a. The spleen is located in the lower left quadrant of the abdomen.
    b. The spleen functions in the body's defense against infection and as a reservoir for blood.
    c. The structure of the spleen is exactly like that of a lymph node.
    d. Splenic pulp contains large phagocytes on the lining of its venous sinuses.

14. Agents that enter the body and cause disease are called _____ .

15. The skin is an example of which of the following defense mechanisms?
    a. immunity
    b. inflammation
    c. mechanical barrier
    d. phagocytosis

16. Which of the following characteristics of the stomach enable it to act as a defense mechanism?
    a. low pH
    b. presence of lysozyme
    c. presence of amylase
    d. presence of pepsin

17. List the major symptoms of inflammation. _____

18. Phagocytes that remain fixed in position within various organs are called
    a. neutrophils.
    b. monocytes.
    c. macrophages.

19. Macrophages are located in the lining of blood vessels in the bone marrow, liver, spleen, and lymph nodes; they form the _____ system.

20. The resistance to specific foreign agents in which certain cells recognize the foreign substances and act to destroy them is _____ .

21. Some undifferentiated lymphocytes migrate to the _____ _____ where they undergo changes and are then called T-lymphocytes.

22. Foreign proteins to which lymphocytes respond are called _____ .

23. Lymphocytes seem to be able to recognize specific foreign proteins because
    a. of changes in the nucleus of the lymphocyte.
    b. the cytoplasm of the lymphocyte is altered.
    c. there are changes in the permeability of the cell membrane of the lymphocyte.
    d. of the presence of receptor molecules on lymphocytes, which fit the molecules of antigens.

24. B-lymphocytes respond to foreign protein by
    a. phagocytosis.
    b. interacting directly with pathogens.
    c. producing antigens.
    d. producing antibodies.

25. In which of the following ways are primary and secondary immune responses different?
    a. Primary responses are more important than secondary responses.
    b. Primary responses produce more antibodies than secondary responses.
    c. A primary response is a direct response to an antigen; a secondary response is indirect.
    d. A primary response is the initial response to an antigen; a secondary response is all subsequent responses to that antigen.

26. A person who receives ready-made antibodies develops artificially acquired _____ immunity.

27. An individual with special abilities to carry on abnormal immune reactions against usually harmless substances is said to have
    a. an increased potential for cancer.
    b. an allergic reaction.
    c. a collagen disease.

28. A common problem following organ transport is _____ _____ .

Study Activities

I. Aids to Understanding Words

Define the following word parts. (p. 429)

gen-

humor-

immun-

inflamm-

nod-

patho-

II. Introduction and Lymphatic Pathways (pp. 430–32)

A. Describe the relationship between the lymphatic system and the cardiovascular system. (p. 430)

B. What are the components of a lymphatic pathway? (p. 430)

III. Tissue Fluid and Lymph (pp. 432–33)

A.   How is tissue fluid formed? Include the composition of tissue fluid and transport mechanisms used. (p. 433)

B.   How is lymph formed, and how is lymph formation related to tissue fluid formation? (p. 433)

C.   What is the function of lymph? (p. 433)

IV. Movement of Lymph (p. 433)

Describe the forces responsible for the circulation of lymph. (pp. 433–34)

V. Lymph Nodes (pp. 434–35)

A.   Label these structures in the accompanying drawing: afferent vessel, sinus, nodule, efferent vessel, hilum, capsule. In addition, indicate the direction of lymph flow through the lymph node. (p. 434)

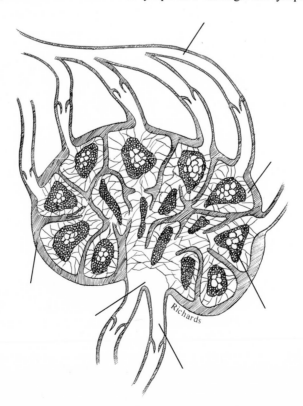

B.   What is the function of lymph nodes? (p. 435)

VI. Thymus and Spleen (pp. 435–36)

A.   Describe the location and function of the thymus gland. (p. 435)

B.   Answer these questions concerning the spleen. (pp. 435–36)
1.   Where is the spleen located?

2.   Label these structures in the accompanying diagram: spleen, capsule, venous sinuses, vein, artery, red pulp, white pulp, splenic artery and vein.

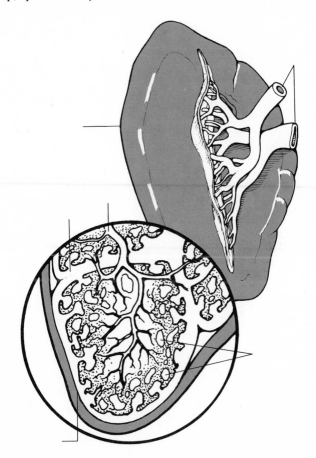

3.   What characteristics of the spleen allow it to function as a blood reservoir?

4.   What characteristics of the spleen allow it to function in the defense against foreign particles?

VII. Body Defenses against Infection (p. 436)

    A.   What kinds of agents cause disease? (p. 436)

    B.   What two major types of defense mechanisms prevent disease? (p. 436)

VIII. Nonspecific Resistance (p. 437)

    A.   What is species resistance? (p. 437)

    B.   What structures function as mechanical barriers? (p. 437)

    C.   What enzymes help us resist infection? (p. 437)

    D.   What is interferon, and how does it work? (p. 437)

    E.   Answer these questions concerning inflammation. (p. 437)
       1.   What is inflammation?

       2.   Explain the reason for the major symptoms of inflammation.
          redness

          swelling

          heat

          pain

       3.   Describe how inflammation is a defense against infection.

       4.   How does the function of reticuloendothelial tissue contribute to the body's defense?

IX. Immunity (pp. 437–45)

    A.   Answer these questions concerning immunity and the origin of lymphocytes. (pp. 437–39)
       1.   What is immunity?

       2.   Where do lymphocytes originate?

       3.   Why are some lymphocytes called T-lymphocytes, while others are called B-lymphocytes?

B. Answer these questions concerning the functions of lymphocytes. (p. 439)
   1. What are the functions of lymphocytes?

   2. How do lymphocytes carry out these functions?

   3. Define *cell-mediated immunity* and *antibody-mediated immunity*.

C. Answer these questions concerning primary and secondary immune responses. (pp. 442–43)
   1. Define *primary immune response*.

   2. Define *secondary immune response*.

D. Define these terms concerning types of immunity. (p. 443)
   naturally acquired active immunity

   artificially acquired active immunity

   artificially acquired passive immunity

   naturally acquired passive immunity
E. Describe the events of an allergic reaction. (pp. 444–45)

F. Answer these questions concerning transplantation and tissue rejection. (p. 445)
   1. What is a tissue rejection reaction?

   2. In what two ways can this reaction be minimized?

When you have completed the study activities to your satisfaction, retake the mastery test and compare your performance with your initial attempt. If there are still areas you do not understand, repeat the appropriate study activities.

17 Urinary System

Overview

This chapter is about the urinary system, which plays a vital role in maintenance of the internal environment by excreting nitrogenous waste products and by selectively excreting or retaining water and electrolytes. The chapter identifies, locates, and describes the functions of the organs of the urinary system (objectives 1, 2, and 3). It explains the structure and function of the nephron—the basic unit of function of the kidney (objective 5). It traces the pathway of blood through the blood vessels of the kidney, explains how glomerular filtrate is produced, and discusses the role of tubular reabsorption in urine production (objectives 4, 6, 7, and 8). It also describes the role of tubular secretion in urine formation (objective 9). It discusses the structure of the ureters, urinary bladder, and urethra and how they function in micturition (objectives 10 and 11).

A study of the urinary system is basic to understanding how the body maintains its chemistry within very narrow limits.

Chapter Objectives

After you have studied this chapter, you should be able to

1. Name the organs of the urinary system and list their general functions.
2. Describe the locations of the kidneys and the structure of a kidney.
3. List the functions of the kidneys.
4. Trace the pathway of blood through the major vessels within a kidney.
5. Describe a nephron and explain the functions of its major parts.
6. Explain how glomerular filtrate is produced and describe its composition.
7. Explain how various factors affect the rate of glomerular filtration and how this rate is regulated.
8. Discuss the role of tubular reabsorption in urine formation.
9. Define *tubular secretion* and explain its role in urine formation.
10. Describe the structure of the ureters, urinary bladder, and urethra.
11. Discuss the process of micturition and explain how it is controlled.

Focus Question

What mechanisms are used by the urinary system to meet its responsibilities for excretion of the end products of protein metabolism and for the maintenance of fluid and electrolyte balance?

Mastery Test

Now take the mastery test. Do not guess. As soon as you complete the test, correct it. Note your successes and failures so that you can read the chapter to meet your learning needs.

1. Those organs of the urinary system that function primarily to transport urine are the
   a. kidney.
   b. urethra.
   c. ureters.
   d. bladder.

2. The kidneys are located
   a. within the abdominal cavity.
   b. between the twelfth thoracic and third lumbar vertebrae.
   c. posterior to the parietal peritoneum.
   d. just below the diaphragm.

3. The superior end of the ureters is expanded to form the funnel-shaped _____

_____ .

4. A series of small elevations that project into the renal sinus from the substance of the kidney and form the sinus wall are called
   a. renal pyramids.
   b. the renal medulla.
   c. renal calyces.
   d. renal papillae.

5. The blood supply to the nephron is via
   a. the renal artery.
   b. an interlobar artery.
   c. an arciform artery.
   d. afferent arterioles.

6. The structure of the renal corpuscle consists of the
   a. glomerulus.
   b. Bowman's capsule.
   c. descending loop of Henle.
   d. proximal convoluted tubule.

7. Blood leaves the capillary cluster of the renal capsule via
   a. an afferent arteriole.
   b. an efferent arteriole.
   c. the peritubular capillary system.
   d. the renal vein.

8. The end product of kidney function is _____ .

9. The transport mechanism used in the glomerulus is
   a. filtration.
   b. osmosis.
   c. active transport.
   d. diffusion.

10. The fluid formed in the capillary cluster of the nephron is the same as blood plasma except for the absence of
    a. glucose.
    b. larger molecules of plasma protein.
    c. bicarbonate ions.
    d. creatinine.

11. Blood pressure affects urine formation because _____ _____ of the blood is necessary to the transport mechanism used in the glomerulus.

12. How much fluid filters through the glomerulus in a 24-hour period?
    a. 1½ quarts
    b. 2 cups
    c. 45 gallons
    d. 5–10 quarts

13. Which of the following substances are present in glomerular filtrate but not in urine?
    a. urea
    b. sodium
    c. potassium
    d. glucose

14. Substances such as sodium ions are reabsorbed in the
    a. proximal convoluted tubule.
    b. distal convoluted tubule.
    c. descending limb of the loop of Henle.
    d. ascending limb of the loop of Henle.

15. The permeability of the distal segment of the tubule to water is regulated by
    a. blood pressure.
    b. ADH.
    c. aldosterone.
    d. renin.

16. The mechanism by which greater amounts of a substance may be excreted in urine than were filtered from the plasma in the glomerulus is
    a. tubular absorption.
    b. active transport.
    c. pinocytosis.
    d. tubular secretion.

17. Which of the following substances enter the urine using tubular secretion?
    a. lactic acid
    b. hydrogen ions
    c. amino acids
    d. potassium

18. The normal output of urine for an adult in an hour is
    a. 20–30 cc.
    b. 30–40 cc.
    c. 40–50 cc.
    d. 50–60 cc.

19. Urine is conveyed from the kidney to the bladder via the _____ .

20. Urine moves along the ureters via
    a. hydrostatic pressure.
    b. gravity.
    c. peristalsis.

21. The internal floor of the bladder has three openings in a triangular area called the _____ .

22. The third layer of the bladder is composed of smooth muscle fibers and is called the
    a. micturition muscle.
    b. detrusor muscle.
    c. urinary muscle.
    d. sympathetic muscle.

23. When stretch receptors in the bladder send impulses along parasympathetic paths, the individual experiences a sensation known as _____ .

24. The usual amount of urine voided at one time is about
    a. 50 ml.
    b. 150 ml.
    c. 500 ml.
    d. 1,000 ml.

25. Which of the following structures is under conscious control?
    a. external urethral sphincter
    b. internal urethral sphincter
    c. bladder wall

Study Activities

I. Aids to Understanding Words

Define the following word parts. (p. 451)

calyc-

cort-

detrus-

glom-

nephr-

mict-

papill-

trigon-

II. The Kidney (pp. 452–57)

A.   Identify the parts of the urinary system in the accompanying drawing: kidney, bladder, ureter, aorta, inferior vena cava, hilum, renal artery, renal vein. (p. 452)

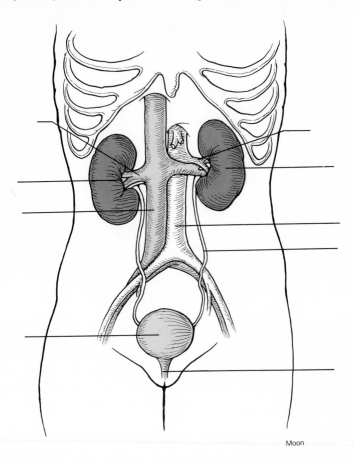

Moon

B.   Describe the precise location of the kidneys. (p. 452)

C. Label these structures in the accompanying drawing: renal capsule, renal pelvis, major and minor calyces, renal cortex, renal medulla, renal papilla, renal pyramid, ureter, renal column. (p. 453)

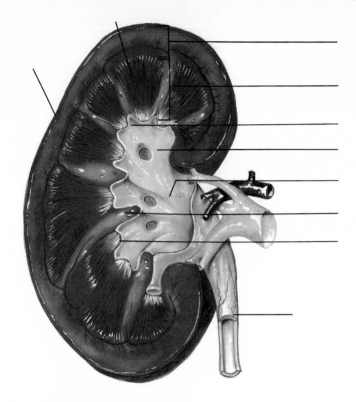

D. What is the functional unit of the kidney? (p. 454)

E. What are the arterial and venous pathways in the kidney? (p. 454)

III. Urine Formation (pp. 457–64)
   A. Answer these questions concerning urine formation. (p. 457)
      1. What are the functions of nephrons?

      2. What mechanisms are used by the nephron to accomplish these functions?

   B. Answer these questions concerning glomerular filtration. (p. 457)
      1. Describe glomerular filtration.

      2. What is the amount and composition of glomerular filtrate?

C. Answer these questions concerning tubular reabsorption. (pp. 460–61)
  1. Explain this statement: Tubular reabsorption is selective. Illustrate it by describing the reabsorption of glucose and amino acids.

  2. Define *renal threshold*.

  3. List the substances reabsorbed by the epithelium of the proximal convoluted tubule.

  4. Describe water and sodium reabsorption in the proximal segment of the renal tubule.

  5. Describe exactly how ADH affects the nephron.

  6. How is urea reabsorbed?

D. Answer these questions concerning tubular secretion. (p. 463)
  1. What is tubular secretion?

  2. Where are hydrogen ions secreted?

  3. How are potassium ions secreted?

E. Answer these questions concerning the composition of urine. (p. 464)
  1. What is the composition of normal urine?

  2. What is the normal output of urine?

IV. Elimination of Urine (p. 464)
  A. Answer these questions concerning the ureters. (pp. 464–65)
    1. Describe the location and structure of the ureters.

    2. How is urine moved through the ureters?

B. Answer these questions concerning the urinary bladder. (pp. 465–66)
    1. How does the bladder change as it fills with urine?

    2. How is urine prevented from spilling back into the ureters?

    3. Describe the structure of the bladder wall.
C. Describe the process of micturition. Be sure to include both autonomic and voluntary events. (p. 466)

When you have completed the study activities to your satisfaction, retake the mastery test and compare your performance with your initial attempt. If there are still areas you do not understand, repeat the appropriate study activities.

18 Water and Electrolyte Balance

Overview

This chapter presents the roles of several body systems in maintaining proper concentrations of water and electrolytes. It defines water and electrolyte balance and explains its importance (objective 1). It tells how water enters the body, how water is distributed in various body compartments, and how it leaves the body (objectives 2 and 3). It explains how electrolytes enter and leave the body and how their concentration is regulated (objective 4). It also describes acid-base balance and mechanisms regulating this balance (objectives 5–7).

Water and electrolyte balance affects and is affected by the various chemical interactions of the body. A knowledge of the mechanisms that control fluid and electrolyte concentrations is essential to understanding the nature of the internal environment.

Chapter Objectives

After you have studied this chapter, you should be able to

1. Explain what is meant by water and electrolyte balance and discuss the importance of this balance.
2. Describe how body fluids are distributed within compartments, how the fluid composition differs between compartments, and how fluids move from one compartment to another.
3. List the routes by which water enters and leaves the body and explain how water input and output are regulated.
4. Explain how electrolytes enter and leave the body and how the input and output of electrolytes are regulated.
5. List the major sources of hydrogen ions in the body.
6. Distinguish between strong and weak acids and bases.
7. Explain how changing pH values of body fluids are minimized by chemical buffer systems, the respiratory center, and the kidneys.

Focus Question

The actions of which body systems must be coordinated in order to maintain normal concentrations of fluids and electrolytes?

Mastery Test

Now take the mastery test. Do not guess. As soon as you complete the test, correct it. Note your successes and failures so that you can read the chapter to meet your learning needs.

1. *Fluid and electrolyte balance* implies that the quantities of these substances entering the body _____ the quantities leaving the body.

2. Which of the following statements about fluid and electrolyte balance is (are) true?
   a. Fluid balance is independent of electrolyte balance.
   b. The concentration of an individual electrolyte is the same throughout the body.
   c. Water and electrolytes occur in compartments in which the composition of fluid varies.
   d. Water is evenly distributed throughout the tissues of the body.

3. More than 60% of body fluid occurs within the cells in a compartment called the _____ fluid compartment.

4. Blood and cerebrospinal fluid occur in the _____ fluid compartment.

5. Which of the following electrolytes is (are) most concentrated within the cell?
   a. sodium
   b. bicarbonate
   c. chloride
   d. potassium

6. Plasma leaves the capillary at the arteriole end and enters interstitial spaces because of
   _____ pressure.

7. Fluid returns from the interstitial spaces to the plasma at the venule end of the capillary because of
   _____ pressure.

8. The most important source of water for the normal adult is
   a. in the form of beverages.
   b. from moist food, such as lettuce and tomatoes.
   c. from oxidative metabolism of nutrients.

9. Thirst is experienced when
   a. the mucosa of the mouth begins to lose water.
   b. salt concentration in the cell increases.
   c. the hypothalamus is stimulated by increasing osmotic pressure of extracellular fluid.
   d. the cortex of the brain is stimulated by shifts in the concentration of sodium.

10. The primary regulator of water output is
    a. loss in the feces.
    b. evaporation as sweat.
    c. urine production.
    d. loss with respiration.

11. The amount of water lost or retained is regulated by
    a. changes in metabolic rate.
    b. an increase or decrease in respiratory rate.
    c. the amount and consistency of feces.
    d. an antidiuretic hormone.

12. The primary source of electrolytes is _____ and _____ .

13. List three routes by which electrolytes are lost.

14. Sodium and potassium ion concentrations are regulated by the kidneys and the hormone
    _____ .

15. Calcium ion concentration in the plasma is regulated by _____ _____ .

16. Acid-base balance is mainly concerned with regulating _____
    _____ concentration.

17. Anaerobic respiration of glucose produces
    a. carbonic acid.
    b. lactic acid.
    c. acetoacetic acid.
    d. ketones.

18. The strength of an acid depends on
    a. the number of hydrogen ions in each molecule.
    b. the nature of the inorganic salt.
    c. the degree to which molecules ionize in water.
    d. the concentration of acid molecules.

19. A base is a substance that will _____ _____ hydrogen ions.

20. A buffer is a substance that
    a. returns an acid solution to neutral.
    b. converts acid solutions to alkaline solutions.
    c. converts strong acids or bases to weak acids or bases.
    d. returns an alkaline solution to neutral.

21. The most important buffer system in plasma and intracellular fluid is the
    a. bicarbonate buffer.
    b. phosphate buffer.
    c. protein buffer.

22. The respiratory center controls hydrogen ion concentration by controlling the _____ and
    _____ of respirations.

23. The slowest acting of the mechanisms that control pH is the
    a. buffer systems.
    b. respiratory system.
    c. kidneys.

Study Activities

I. Aids to Understanding Words

Define the following word parts. (p. 473)

de-   intra-

extra-   neutr-

im- (or in-)

II. Introduction and Distribution of Body Fluids (pp. 474–77)

A. Because electrolytes are dissolved in water, water and electrolyte balance are _____ .
    (p. 474)

B. Answer these questions concerning the composition of body fluids. (pp. 475–76)
    1. What is the composition of extracellular fluid?

    2. What is the composition of intracellular fluid?

C. What mechanisms are responsible for the movement of fluid and electrolytes from one compartment to
    another? Describe these mechanisms as fully as possible. (pp. 476–77)

III. Water Balance (pp. 477–78)

A. Answer these questions concerning water intake. (p. 477)
    1. Water balance exists when gain of water from all sources _____ the total loss
        of water.
    2. From what sources is this water derived?

B. Describe the mechanism that regulates the intake of water. (p. 477)

C. By what routes is water lost from the body? (p. 478)

D. The hormone that regulates water balance is _____ . (p. 478)

IV. Electrolyte Balance (pp. 478–80)

   A.   Answer these questions concerning electrolyte intake. (p. 478)
      1.   List the electrolytes that are important to cellular function.

      2.   What are the sources of these electrolytes?

   B.   List three routes by which electrolytes are lost. (p. 479)

   C.   Answer these questions concerning regulation of electrolyte balance. (pp. 479–80)
      1.   What is the role of aldosterone in electrolyte regulation?

      2.   What is the role of parathyroid hormone in electrolyte regulation?

      3.   What is the role of the kidney in electrolyte regulation?

      4.   How is the concentration of negatively charged ions regulated?

V. Acid-Base Balance (pp. 480–83)

   A.   Answer these questions concerning acid-base balance. (p. 480)
      1.   What is an acid?

      2.   What is a base?

      3.   What is acid-base balance?

   B.   List and describe the sources of hydrogen ions in the body. (p. 480)

   C.   What is the difference between a strong acid or base and a weak acid or base? (p. 481)

D. Answer these questions concerning regulation of hydrogen ion concentration. (pp. 481–83)
    1. How is hydrogen ion concentration regulated by acid-base buffer systems?

    2. Describe how this works with the bicarbonate buffer system, the phosphate buffer system, and the protein buffer system.

    3. How does the respiratory center regulate hydrogen ion concentration?

    4. How do the kidneys help regulate hydrogen ion concentration?

    5. How do these mechanisms differ from each other, especially with respect to speed of action?

When you have completed the study activities to your satisfaction, retake the mastery test and compare your performance with your initial attempt. If there are still areas you do not understand, repeat the appropriate study activities.

19 Reproductive Systems

Overview

This chapter explains reproductive systems; unique systems because they are essential to the survival of the species rather than to the survival of the individual. The chapter explains the structure and functions of the male and female reproductive systems (objectives 1, 2, 3, 5, 6, 7, and 10). It also describes the hormonal events that control the male and female reproductive systems, as well as the major events of the menstrual cycle (objectives 4, 8, and 9).

Understanding the processes of sexual function contributes to an understanding of humans as sexual beings.

Chapter Objectives

After you have studied this chapter, you should be able to

1. Name the parts of the male reproductive system and describe the general functions of each part.
2. Describe the structure of a testis and explain how sperm cells are formed.
3. Trace the path followed by sperm cells from the site of their formation to the outside.
4. Explain how hormones control the activities of the male reproductive organs.
5. Name the parts of the female reproductive system and describe the general functions of each part.
6. Describe the structure of an ovary and explain how egg cells and follicles are formed.
7. Trace the path followed by an egg cell after ovulation.
8. Describe how hormones control the activities of the female reproductive organs.
9. Describe the major events that occur during a menstrual cycle.
10. Review the structure of the mammary glands.

Focus Question

How do the male and female reproductive systems differ and complement each other's function?

Mastery Test

Now take the mastery test. Do not guess. As soon as you complete the test, correct it. Note your successes and failures so that you can read the chapter to meet your learning needs.

1. The primary sex organs (gonads) of the male reproductive system are the _____ .

2. Male sex cells are produced by the _____ _____ .

3. Prior to adolescence, the undifferentiated spermatogenic cells in the testes are called

   _____ .

4. Sex cells are produced in a process called _____ .

5. How many chromosomes does each spermatogonium contain?

6. During meiosis the chromosome number is
   a. reduced.   c. unchanged.
   b. increased.

7. The nucleus in the head of the sperm contains _____ chromosomes.

8. The function of the epididymis is
   a. to produce sex hormones.   c. to store sperm as they mature.
   b. to provide the sperm with mobile tails.   d. to supply some of the force needed for ejaculation.

9. Which of the following substances are added to sperm cells by the seminal vesicle?
   a. acid
   b. fructose
   c. glucose

10. The function of the secretion of the bulbourethral glands is to
    a. neutralize the acid secretions of the vagina.
    b. nourish sperm cells.
    c. lubricate the penis.
    d. increase the volume of seminal fluid.

11. Which of the following is not a male internal accessory organ?
    a. epididymis
    b. vas deferens
    c. testes
    d. seminal vesicle

12. The external organs of the male reproductive system include the
    a. penis.
    b. testes.
    c. prostate gland.
    d. scrotum.

13. Erection of the penis depends on
    a. contraction of the perineal muscles.
    b. filling of the corpus spongiosum with arterial blood.
    c. enlargement of the glans penis.
    d. peristaltic contractions of the vas deferens.

14. Hormones that control male reproductive functions are secreted from the _____ , the _____ , and the _____ _____ _____ .

15. The pituitary hormone that stimulates the testes to produce testosterone is
    a. gonadotropin releasing hormone.
    b. FSH.
    c. LH (ICSH).
    d. ACTH.

16. In the male, growth of body hair, especially in the axilla, face, and pubis, and increased muscle and bone development are examples of _____ _____ characteristics.

17. The hormone which provokes these changes is _____ .

18. The primary sex organs (gonads) of the female reproductive system are the _____ .

19. In the ovary, the primary germinal epithelium is located
    a. in the medulla.
    b. between the medulla and cortex.
    c. in the cortex.
    d. on the free surface of the ovary.

20. How many mature egg cells are produced by each primary oocyte? _____

21. At puberty, the primary oocyte matures within the _____ _____ .

22. The egg is released from the ovary in a process called _____ .

23. Which of the following statements is (are) true about the fallopian (uterine) tubes?
    a. The end of the uterine tube near the ovary has many fingerlike projections called fimbriae.
    b. The fimbriae are attached to the ovaries.
    c. The inner layer of the ovarian tube is cuboidal epithelium.
    d. There are cilia in the lining of the fallopian tube that help move the egg toward the uterus.

24. The inner layer of the uterus is the _____ .

25. Which of the following statements about the vagina is (are) true?
    a. The mucosal layer contains many mucous glands.
    b. The bulbospongiosus muscle is primarily responsible for closing the vaginal orifice.
    c. The vagina connects the uterus to the outer surface of the body.
    d. The hymen is a membrane that covers the mouth of the cervix.

26. The organ of the female reproductive system that corresponds to the penis is the
    a. vagina.
    b. mons pubis.
    c. clitoris.
    d. labia majora.

27. Which of the following tissues become engorged and erect in response to sexual stimulation?
    a. clitoris
    b. labia minora
    c. outer ⅓ of vagina
    d. upper ⅓ of vagina

28. The hormonal mechanisms that control female reproductive functions are (more, less) complex than in the male.

29. The primary female sex hormones are _____ and _____ .

30. Which of the following secondary sex characteristics in the female seem to be related to androgen concentration?
    a. breast development
    b. growth of axillary and pubic hair
    c. female skeletal configuration
    d. deposition of adipose tissue over hips, thighs, buttocks, and breasts

31. During the menstrual cycle, the event that seems to provoke ovulation is
    a. increasing levels of progesterone.
    b. sudden increase in concentration of LH.
    c. decreasing levels of estrogen.
    d. cessation of secretion of FSH.

32. After release of the egg, the follicle forms a(n) _____ _____ .

33. As the above structure develops, the level of which of the following hormones increases?
    a. estrogen
    b. progesterone
    c. FSH
    d. LH

34. As the hormone levels change in that part of the cycle before and immediately after ovulation, which of the following changes are seen in the uterus?
    a. growth of the myometrium
    b. increase in adipose cells of the perimetrium
    c. thickening of the endometrium
    d. decrease in uterine gland activity

35. The concentration of which of the following hormones decreases following ovulation?
    a. estrogen
    b. progesterone
    c. FSH
    d. LH

36. The cessation of the menstrual cycle in middle age is called _____ .

Study Activities

I. Aids to Understanding Words

Define the following word parts. (p. 491)

andr-   germ-

ejacul-   labi-

fimb-   mens-

follic-   mons-

genesis   puber-

II. Introduction and Organs of the Male Reproductive System (pp. 492–96)

A. Describe the functions of the male and female reproductive systems. (p. 492)

B. Answer these questions concerning the organs of the male reproductive systems. (p. 492)
   1. Label these structures in the accompanying diagram: urinary bladder, vas deferens, prostate gland, penis, urethra, prepuce, glans penis, epididymis, testis, scrotum, corpus cavernosum, corpus spongiosum, bulbourethral gland, ejaculatory duct, seminal vesicle, symphysis pubis, anus.

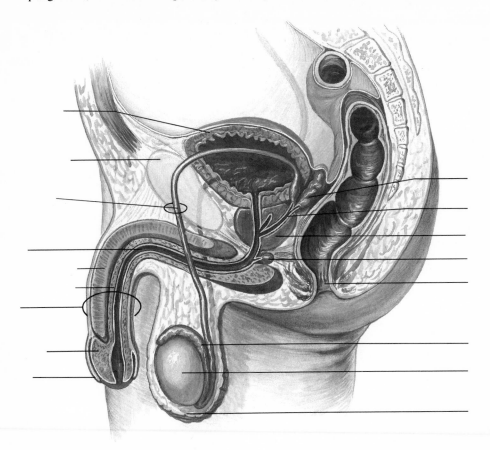

   2. What is the primary function of the male reproductive system?

   3. What are the primary organs of the male reproductive system?

C. Answer these questions concerning the structure of the testes. (p. 493)
   1. What is the function of germinal epithelium?

   2. What is the function of interstitial cells?

D. Describe the process of sperm formation, including where sperm are stored and a description of the sperm. (p. 494)

E. Answer these questions concerning meiosis. (p. 494)
    1. What cells undergo meiosis?

    2. What is the result of meiosis in the male?

III. Male Internal Accessory Organs (pp. 494–96)

A. List the internal accessory organs of the male reproductive system. (p. 494)

B. Answer these questions concerning the epididymis. (pp. 494–95)
    1. Describe the location and structure of the epididymis.

    2. What is the function of the epididymis in the emission of sperm?

C. Describe the course of the vas deferens. (p. 495)

D. Answer these questions concerning the seminal vesicles. (p. 495)
    1. Where are the seminal vesicles located?

    2. What is the nature and function of the secretion of the seminal vesicles?

E. Answer these questions concerning the prostate gland. (p. 496)
    1. Where is the prostate gland located?

    2. What is the nature and function of the secretion of the prostate gland?

    3. When is this secretion released?

F. Answer these questions concerning the bulbourethral glands. (p. 496)
   1. Describe the location and function of the bulbourethral glands (Cowper's glands).

   2. What is the stimulus for release of this secretion?

G. Describe the seminal fluid. (p. 496)

IV. Male External Reproductive Organs and Erection, Orgasm, and Ejaculation (pp. 497–98)

A. Describe the structure of the penis. (p. 497)

B. Describe the events of erection and ejaculation. (pp. 497–98)

V. Hormonal Control of Male Reproductive Functions (pp. 498–99)

A. Answer these questions concerning pituitary hormones. (p. 498)
   1. What seems to initiate the changes of puberty?

   2. What are the functions of FSH and LH?

B. Answer these questions concerning male sex hormones. (pp. 498–99)
   1. Where is testosterone produced?

   2. What is the function of testosterone?

   3. List the male secondary sexual characteristics.

C. Describe the regulation of sex hormones in the male. (p. 499)

VI. Organs of the Female Reproductive System (pp. 500–503)

A. Label these structures in the accompanying drawing: ovary, uterine tube, uterus, vagina, anus, urinary bladder, urethra, clitoris, labium minor, labium major, symphysis pubis, fimbriae cervix, rectum, vaginal orifice. (p. 500)

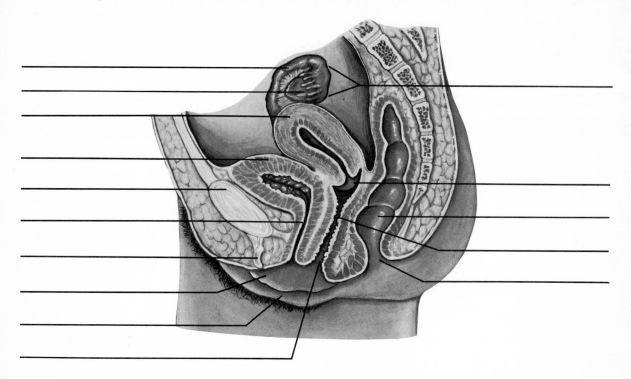

B. Describe the structure of the ovaries. (p. 500)

C. Answer these questions concerning the formation of egg cells. (p. 501)
1. Outline the process of egg cell production.

2. How is this different from spermatogenesis?

D. Answer these questions concerning the maturation of a follicle. (pp. 501–2)
1. What stimulates maturation of a primary follicle at puberty?

2. What changes occur in the follicle as a result of maturation?

E. Answer these questions concerning ovulation. (pp. 502–3)

    1. What provokes ovulation?

    2. What happens to the egg after it leaves the ovary?

VII. Female Internal Accessory Organs (pp. 503–4)

A. Label these structures in the accompanying drawing: body of uterus, uterine tube, infundibulum, ovaries, cervix, vagina, cervical orifice, fimbriae, egg cell, follicle, endometrium, myometrium, perimetrium, broad ligament, round ligament. (p. 503)

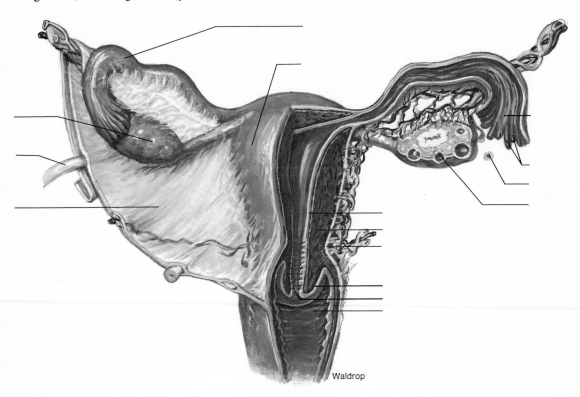

Waldrop

B. How does the structure of the uterine tube move the egg toward the uterus? (p. 503)

C. Answer these questions concerning the uterus. (p. 504)
   1. Based on the description in the text, draw a uterus. Locate the body, cervix, endometrium, myometrium, and perimetrium.

   2. How are the muscle fibers arranged in the myometrium?

D. Answer these questions concerning the vagina. (p. 504)
   1. What is the function of the vagina?

   2. Describe the structure of the vagina.

VIII. Female External Reproductive Organs and Erection, Lubrication, and Orgasm (pp. 504–5)

A. To what male organ do the labia majora correspond? (p. 504)
B. Describe the labia minora. (p. 504)

C. What is the structure of the clitoris? (p. 505)

D. Answer these questions concerning the vestibule. (p. 505)
   1. Describe the location of the vestibule.

   2. What is the function of the vestibular glands?

E.   Describe the events of erection and orgasm in the female. (p. 505)

IX. **Hormonal Control of Female Reproductive Functions** (pp. 506–8)

A.   Answer these questions concerning female sex hormones. (p. 506)
1.   What appears to initiate sexual maturation in the female?

2.   What are the sources of female sex hormones?

3.   What is the function of estrogen?

4.   What is the function of progesterone?

B.   Answer these questions concerning female reproductive cycles. (pp. 506–8)
1.   How is a female's first menstrual cycle initiated?

2.   Describe the events of the menstrual cycle. Include shifts in hormone levels, uterine changes, and ovarian changes.

C.   Answer these questions concerning menopause. (p. 508)
1.   What is menopause?

2.   What seems to be the cause of menopause?

X. **Mammary Glands** (p. 509)
A.   What is the function of the mammary glands? (p. 509)

B.   Where are the mammary glands located? (p. 509)

C.   Describe the structure of the mammary glands. (p. 509)

When you have completed the study activities to your satisfaction, retake the mastery test and compare your performance with your initial attempt. If there are still areas you do not understand, repeat the appropriate study activities.

20 Pregnancy, Growth, and Development

Overview

This chapter describes the events of fertilization and the changes that take place in the maternal body during pregnancy and childbirth (objectives 2, 4, 6, and 9). It also covers the events of embryonic and fetal development and the adjustments of the infant to extrauterine life (objectives 3, 5, 7, 8, and 10). The concepts of growth and development are also covered (objective 1).

Chapter Objectives

After you have studied this chapter, you should be able to

1. Distinguish between growth and development.
2. Define *pregnancy* and describe the process of fertilization.
3. Describe the major events that occur during the period of cleavage.
4. Describe the hormonal changes that occur in the maternal body during pregnancy.
5. Explain how the primary germ layers originate and list the structures produced by each layer.
6. Describe the formation and function of the placenta.
7. Define *fetus* and describe the major events that occur during the fetal stage of development.
8. Trace the general path of blood through the fetal circulatory system.
9. Describe the birth process and explain the role of hormones in the process.
10. Describe the major circulatory and physiological adjustments that occur in the newborn.

Focus Question

How is a unique individual produced from the union of two cells?

Mastery Test

Now take the mastery test. Do not guess. As soon as you complete the test, correct it. Note your successes and failures so that you can read the chapter to meet your learning needs.

1. An increase in size and number of cells is referred to as _____ .

2. Fertilization takes place in the
   a. vagina.
   b. cervix.
   c. uterus.
   d. uterine tube.

3. Which of the following is thought to be the mechanism by which the sperm enters the egg?
   a. An antigen-antibody reaction that briefly alters the cell membrane of the egg.
   b. The structure of the cell membrane of the egg allows entry.
   c. The head of the sperm has an enzyme that permits digestion through the membrane of the egg.
   d. The mechanism is unknown.

4. The first phase in embryonic development is called _____ .

5. Implantation takes place by the end of _____ week(s) following fertilization.

6. The hormone that maintains the corpus luteum following implantation is
   a. LH.
   b. HCG.
   c. progesterone.
   d. FSH.

7. The primary source of hormones needed to support a pregnancy after the first 3 months is the

_____ .

8. Relaxation of the ligaments of the pelvic girdle is due to high concentrations of _____
   _____ .

9. The ectoderm, mesoderm, and endoderm are _____ _____ layers.

10. From which of the layers of the embryonic disk do the hair, nails, and glands of the skin arise?
    a. endoderm   c. mesoderm
    b. ectoderm

11. Which of the following structures arises from the mesoderm?
    a. lining of the mouth   c. lining of the respiratory tract
    b. muscle   d. epidermis

12. At what time does the embryonic disk become a cylinder?
    a. 4 weeks of development   c. 8 weeks of development
    b. 6 weeks of development   d. 4 days of development

13. The chorion in contact with the endometrium becomes the _____ .

14. The membrane covering the embryo is called the _____ .

15. The vessels in the umbilical cord are
    a. one artery, one vein.   c. two arteries, one vein.
    b. one artery, two veins.   d. two arteries, two veins.

16. In the embryo, blood cells are formed in the
    a. amnion.   c. allantois.
    b. placenta.   d. yolk sac.

17. The embryonic stage ends at _____ weeks.

18. Pregnancy can be diagnosed 10 days after fertilization by testing the urine of the woman for which of the
    following substances?
    a. estrogen   c. human chorionic gonadotropin
    b. progesterone   d. amniotic fluid

19. At the beginning of the fetal stage of development (most/few) of the body structures are formed.

20. Skeletal muscles become active in the _____ lunar month.

21. A fetus is full term at the end of the _____ lunar month.

22. Oxygen and nutrient-rich blood reach the fetus from the placenta via the umbilical _____ .

23. The ductus venosus shunts blood around the
    a. liver.   c. pancreas.
    b. spleen.   d. small intestine.

24. The structures that allow blood to avoid the nonfunctioning fetal lungs are the _____
    _____ and the _____ _____ .

25. Labor is initiated by a decrease in _____ levels and secretion of
    _____ by the posterior pituitary gland.

26. Bleeding following expulsion of the afterbirth is controlled by
    a. hormonal mechanisms.   c. contraction of the uterine muscles.
    b. increased fibrinogen levels.   d. sympathetic stimulation of arterioles.

27. Milk production is stimulated by the hormone _____ .

28. Milk is secreted from the breast
    a. as production fills the duct structure.
    b. in response to hormonal stimulation.
    c. in response to mechanical stimulation of the nipple, i.e., sucking.
    d. in response to gonadotropins.

29. The factor that decreases the effort required for an infant to breathe after the first breath is

    _____ .

30. The primary energy source for the newborn is
    a. glucose.
    b. fat.
    c. protein.

31. An infant's urine is (more, less) concentrated than an adult's.

32. Which of the following fetal structures closes as a result of a change in pressure?
    a. ductus venosus
    b. ductus arteriosus
    c. umbilical vessels
    d. foramen ovale

Study Activities

I. Aids to Understanding Words

Define the following word parts. (p. 517)

allant-   morul-

chorio-   nat-

cleav-   troph-

lacun-   umbil-

II. Introduction (p. 518)

Define growth and development.

III. Pregnancy (pp. 518–19)

Describe the developmental events from fertilization to implantation.

IV. Prenatal Period (pp. 520–31)

A. Describe the hormonal changes which occur in the mother during pregnancy. (pp. 520–21)

B. Answer these questions concerning the embryonic stage. (pp. 521–25)
    1. What are the boundaries that define the embryonic stage of development?

    2. What is the embryonic disk?

    3. What are the primary germ layers?

4. List the structures that arise from these primary germ layers.

ectoderm

mesoderm

endoderm

5. Label these structures in the accompanying illustration: uterine wall, placenta, umbilical cord, embryonic blood vessels, chorion, villi, lacunae, maternal portion of placenta, maternal blood vessels.

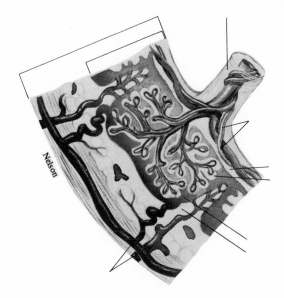

6. Describe the structure and function of the umbilical cord and fetal membranes.

7. What is the function of the allantois and the yolk sac?

C. Fill in the following chart. (pp. 526–28)

The fetal stage

Month	Major Events of Growth and Development
Third Lunar Month	
Fourth Lunar Month	
Fifth Lunar Month	
Sixth Lunar Month	
Seventh Lunar Month	
Eighth Lunar Month	
Ninth Lunar Month	
Tenth Lunar Month	

D. Trace a drop of fetal blood from the placenta through the circulatory system of the fetus. Identify the difference from postnatal circulation. (pp. 528–30)

E. Answer these questions concerning the birth process. (pp. 530–31)
    1. What changes take place in placental hormone secretion after the seventh month of gestation?

    2. What stimulates secretion of oxytocin?

    3. What regulates secretion of oxytocin?

    4. What is labor?

    5. What is afterbirth?

    6. How is bleeding controlled once the uterus is emptied?

V. Postnatal Period (pp. 532–34)

A. Describe the production and secretion of milk. (p. 532)

B. Fill in the following chart. (pp. 532–34)

The neonatal period

Life Function	Major Changes Needed to Adjust to Extrauterine Life
Respiration	
Nutrition	
Urine Formation	
Temperature Control	
Circulation	

When you have completed the study guide to your satisfaction, retake the mastery test and compare your performance to your initial attempt. If there are still areas you do not understand, repeat the appropriate study activities.

Mastery Test Answers

1 Mastery Test Answers

1. c
2. c
3. Latin and Greek
4. anatomy
5. physiology
6. always
7. movement, responsiveness, growth, reproduction, respiration, digestion, absorption, circulation, assimilation, excretion
8. water
9. energy, living matter
10. energy
11. increases
12. hydrostatic
13. a
14. cell, tissue, organ, organ system
15. axial portion
16. appendicular
17. dorsal, ventral
18. diaphragm
19. mediastinum
20. a, b, c
21. pleural cavity
22. pericardial
23. abdominopelvic
24. d
25. a-3, b-4, c-2, d-2, e-4, f-2, g-2, h-3, i-2, j-1
26. b, c
27. c
28. b
29. body regions

2 Mastery Test Answers

1. chemistry
2. Matter is anything that has weight and takes up space. It occurs in the forms of solids, liquids, and gases.
3. elements
4. carbon, hydrogen, oxygen, and nitrogen
5. e
6. a-3, b-1, c-2
7. protons
8. a
9. the outer shell of its atoms has its maximum number of electrons.
10. a
11. a
12. a
13. different
14. molecular
15. structural
16. synthesis, decomposition
17. reversible
18. acid
19. acid
20. 7
21. a. I, b. O, c. O, d. I, e. O, f. O
22. lipids
23. a

3 Mastery Test Answers

1. shape
2. a, b
3. cytoplasm, nucleus
4. b
5. selectively permeable
6. b
7. endoplasmic reticulum
8. true
9. outside
10. energy
11. b
12. d
13. b
14. movement
15. nucleolus, chromatin
16. diffusion
17. facilitated diffusion
18. osmosis
19. a
20. filtration
21. active transport
22. pinocytosis
23. phagocytosis
24. b
25. mitosis
26. a-2, b-4, c-1, d-3
27. differentiation

4 Mastery Test Answers

1. anabolic metabolism
2. catabolic metabolism
3. a
4. c
5. protein
6. hydrolysis, digestion
7. c
8. catalyst
9. b
10. b
11. coenzyme
12. a
13. oxidation
14. anaerobic respiration
15. aerobic respiration
16. oxygen
17. in ATP molecules
18. metabolic pathway
19. monosaccharide
20. more
21. deamination
22. d
23. a, c
24. DNA (deoxyribonucleic acid)
25. chromatin or nucleus, cytoplasm
26. messenger, transfer
27. c
28. mutation

5 Mastery Test Answers

1. similar
2. b
3. a, c
4. a-4, b-1, c-5, d-2, e-3
5. transitional epithelium
6. exocrine
7. a, b, d
8. collagen
9. elastin
10. b
11. a, b, c, d
12. muscle, bone; bone, bone; fibrous
13. bone
14. plasma
15. a
16. smooth, skeletal, cardiac
17. nervous

6 Mastery Test Answers

1. serous, mucous, cutaneous, synovial
2. c
3. b
4. c
5. epidermis
6. dermis
7. subcutaneous layer
8. c
9. a
10. dermis
11. heat insulator
12. c
13. c
14. b
15. c

7 Mastery Test Answers

1. b
2. compact
3. spongy or cancellous
4. marrow
5. intramembranous bones
6. endochondral bones
7. d
8. levers
9. a, c
10. b
11. 206
12. skull, hyoid, vertebral column, thoracic cage
13. pectoral girdle, arms or upper limbs, pelvic girdle, legs or lower limbs
14. b
15. occipital
16. maxillary
17. fontanels
18. b
19. c
20. a, b, c, d
21. thoracic vertebrae, sternum
22. clavicles, scapulae
23. radius
24. a
25. a
26. c
27. b
28. c
29. b
30. a
31. b

8 Mastery Test Answers

1. fascia
2. a, c
3. b
4. actin, myosin
5. c
6. a
7. cholinesterase
8. adenosine triphosphate or ATP
9. creatine phosphate
10. a, c, d
11. lactic acid
12. threshold stimulus
13. c
14. a
15. a
16. more slowly
17. multiunit, visceral
18. a, c
19. rapidly
20. origin, insertion
21. antagonists
22. d
23. a
24. b
25. linea alba
26. b

9 Mastery Test Answers

1. sensory, integrative, motor
2. neuron
3. d
4. yes
5. c
6. b
7. Potassium
8. sodium
9. permeability
10. a
11. synapse
12. facilitation
13. c
14. a
15. sensory, interneuron, motor
16. nerve
17. reflex
18. brain, spinal cord
19. a
20. c
21. c
22. b, d
23. cerebrum, brain stem, cerebellum
24. a
25. a-3, b-4, c-3, d-1, e-2, f-1
26. left
27. choroid plexuses
28. d
29. b
30. limbic system
31. reticular formation
32. somatic, autonomic
33. 12, brain stem
34. 31
35. autonomic
36. thorax, sacrum
37. a, b
38. c, d

10 Mastery Test Answers

1. a, b, c, d, e
2. c
3. a
4. chemicals
5. a, c
6. referred
7. b
8. c
9. olfactory nerve or tracts
10. c
11. b
12. sweet, salt, sour, bitter
13. equilibrium
14. c, d
15. eustachian tube
16. osseous labyrinth, membranous labyrinth
17. perilymph
18. a
19. b
20. c
21. cornea
22. a
23. b
24. optic nerve
25. a
26. cornea
27. pupil
28. retina
29. b
30. refraction
31. rods, cones
32. a-1, b-2, c-1, d-2
33. rhodopsin, opsin, retinal
34. a

11 Mastery Test Answers

1. endocrine
2. a
3. prostaglandins
4. a, b
5. hypothalamus
6. a, b
7. posterior
8. anterior
9. a, b, c
10. prolactin
11. a, d
12. c
13. thyroxine, triiodothyronine
14. c, d
15. iodine
16. calcitonin
17. a, b, c
18. a, d
19. epinephrine, norepinephrine
20. a
21. a, b, c
22. male
23. islets of Langerhans
24. glucagon
25. b, d
26. b

12 Mastery Test Answers

1. alimentary canal
2. accessory organs
3. mixing, propelling
4. no
5. frenulum
6. c
7. c
8. d
9. increase
10. b
11. peristalsis
12. b
13. c
14. b
15. vitamin B_{12}
16. c
17. inhibits
18. chyme
19. a
20. a
21. a
22. b
23. alkaline
24. upper right
25. c
26. bile salts
27. a, d
28. a, b, d
29. duodenum, jejunum, ileum
30. b
31. d
32. cecum
33. b
34. electrolytes, water
35. essential nutrients
36. b, d
37. b, d
38. d
39. free fatty acids, glycerol
40. linoleic acid
41. a
42. cholesterol
43. a, c, d
44. amino acids
45. complete
46. yes
47. a, c
48. fat
49. sunlight or ultraviolet rays
50. b
51. a
52. c
53. calcium, phosphorus
54. c
55. sodium
56. a
57. c

13 Mastery Test Answers

1. a-1, b-2, c-2, d-1, e-3
2. nasal cavity, larynx
3. b, c
4. a
5. c
6. alveolar ducts
7. visceral pleura
8. parietal pleura
9. contracts, increasing, decreasing
10. c
11. pleural
12. surfactant

13. c
14. b
15. d
16. brain stem
17. a
18. carbon dioxide
19. alveolus, capillary
20. pressure
21. partial pressure
22. hemoglobin
23. c

14 Mastery Test Answers

1. plasma
2. 55
3. b
4. a
5. b
6. erythropoietin
7. yes
8. a
9. a
10. b
11. 5,000, 10,000
12. platelet or thrombocyte

13. a-3, b-1, c-2, d-2, e-3
14. b, d
15. b
16. fibrinogen, fibrin
17. a, c
18. d
19. positive feedback
20. embolus
21. b
22. a
23. c

15 Mastery Test Answers

1. thorax
2. a
3. c
4. atria, ventricles
5. a, d
6. b
7. coronary arteries
8. cardiac cycle
9. b
10. a
11. electrocardiogram
12. a
13. decrease
14. b, c
15. c
16. vasoconstriction
17. b, d

18. filtration, osmosis, diffusion
19. a
20. a
21. b
22. valves
23. d
24. left atrium
25. b, c
26. hepatic portal system
27. b
28. stroke volume
29. heart action, blood volume, viscosity, peripheral resistance
30. c
31. parasympathetic
32. b
33. a

16 Mastery Test Answers

1. lymphatic capillaries, collecting ducts
2. b
3. b
4. c
5. a, c
6. veins
7. c
8. d
9. b
10. b
11. thymosin, lymphatic
12. spleen
13. b, d
14. pathogens
15. c
16. a, d
17. redness, swelling, heat, pain
18. c
19. reticuloendothelial
20. immunity
21. thymus gland
22. antigens
23. d
24. d
25. d
26. passive
27. b
28. tissue rejection reaction

17 Mastery Test Answers

1. b, c
2. a, b, c
3. renal pelvis
4. d
5. d
6. a, b
7. b
8. urine
9. a
10. b
11. hydrostatic pressure
12. c
13. d
14. a
15. b
16. d
17. b, d
18. d
19. ureters
20. c
21. trigone
22. b
23. urgency
24. b
25. a

18 Mastery Test Answers

1. equal
2. c
3. intracellular
4. extracellular
5. d
6. hydrostatic
7. osmotic
8. a
9. c
10. c
11. d
12. food, beverages
13. perspiration, feces, urine
14. aldosterone
15. parathyroid hormone
16. hydrogen ion
17. b
18. c
19. combine with
20. c
21. c
22. rate, depth
23. c

19 Mastery Test Answers

1. testes
2. germinal epithelium
3. spermatogonia
4. meiosis
5. 46
6. a
7. 23
8. c
9. c
10. c
11. c
12. a, d
13. b
14. testes, hypothalmus, anterior pituitary gland
15. c
16. secondary sexual
17. testosterone
18. ovaries
19. d
20. 1
21. primary follicle
22. ovulation
23. a, d
24. endometrium
25. b, c
26. c
27. a, c
28. more
29. estrogen, progesterone
30. b, c
31. b
32. corpus luteum
33. b
34. c
35. c
36. menopause

20 Mastery Test Answers

1. growth
2. d
3. d
4. cleavage
5. one
6. b
7. placenta
8. placenta estrogen
9. primary germ
10. b
11. b
12. a
13. placenta
14. amnion
15. c
16. c, d
17. 8
18. c
19. most
20. 5th
21. 10th
22. vein
23. a
24. foramen ovale, ductus arteriosus
25. progesterone, oxytocin
26. c
27. prolactin
28. c
29. surfactant
30. b
31. less
32. d

Credits

Chapter 1
Page 6: From John W. Hole, Jr., *Human Anatomy and Physiology*, 4th ed. Copyright © 1987 Wm. C. Brown Publishers, Dubuque, Iowa. All Rights Reserved. Reprinted by permission.

Chapter 3
Page 18: (bottom) From John W. Hole, Jr., *Human Anatomy and Physiology*, 2d ed. Copyright © 1981 Wm. C. Brown Publishers, Dubuque, Iowa. All Rights Reserved. Reprinted by permission. **Page 19:** From John W. Hole, Jr., *Human Anatomy and Physiology*, 4th ed. Copyright © 1987 Wm. C. Brown Publishers, Dubuque, Iowa. All Rights Reserved. Reprinted by permission.

Chapter 7
Page 37: From John W. Hole, Jr., *Human Anatomy and Physiology*, 2d ed. Copyright © 1981 Wm. C. Brown Publishers, Dubuque, Iowa. All Rights Reserved. Reprinted by permission. **Page 40:** From John W. Hole, Jr., *Human Anatomy and Physiology*, 1st ed. Copyright © 1978 Wm. C. Brown Publishers, Dubuque, Iowa. All Rights Reserved. Reprinted by permission. **Page 42:** From John W. Hole, Jr., *Human Anatomy and Physiology*, 4th ed. Copyright © 1987 Wm. C. Brown Publishers, Dubuque, Iowa. All Rights Reserved. Reprinted by permission.

Chapter 8
Page 47: From John W. Hole, Jr., *Human Anatomy and Physiology*, 3d ed. Copyright © 1984 Wm. C. Brown Publishers, Dubuque, Iowa. All Rights Reserved. Reprinted by permission. **Page 49:** From John W. Hole, Jr., *Human Anatomy and Physiology*, 4th ed. Copyright © 1987 Wm. C. Brown Publishers, Dubuque, Iowa. All Rights Reserved. Reprinted by permission.

Chapter 9
Page 55: From John W. Hole, Jr., *Human Anatomy and Physiology*, 3d ed. Copyright © 1984 Wm. C. Brown Publishers, Dubuque, Iowa. All Rights Reserved. Reprinted by permission. **Page 57:** From John W. Hole, Jr., *Human Anatomy and Physiology*, 4th ed. Copyright © 1987 Wm. C. Brown Publishers, Dubuque, Iowa. All Rights Reserved. Reprinted by permission. **Page 59:** From John W. Hole, Jr., *Human Anatomy and Physiology*, 2d ed. Copyright © 1981 Wm. C. Brown Publishers, Dubuque, Iowa. All Rights Reserved. Reprinted by permission. **Page 60:** From John W. Hole, Jr., *Human Anatomy and Physiology*, 3d ed. Copyright © 1984 Wm. C. Brown Publishers, Dubuque, Iowa. All Rights Reserved. Reprinted by permission.

Chapter 10
Page 69: (top) From John W. Hole, Jr., *Human Anatomy and Physiology*, 4th ed. Copyright © 1987 Wm. C. Brown Publishers, Dubuque, Iowa. All Rights Reserved. Reprinted by permission. **Page 69: (bottom)** From John W. Hole, Jr., *Human Anatomy and Physiology*, 4th ed. Copyright © 1987 Wm. C. Brown Publishers, Dubuque, Iowa. All Rights Reserved. Reprinted by permission.

Chapter 12
Page 83: From Kent M. Van De Graaff, *Human Anatomy.* Copyright © 1984 Wm. C. Brown Publishers, Dubuque, Iowa. All Rights Reserved. Reprinted by permission. **Page 85:** From John W. Hole, Jr., *Human Anatomy and Physiology*, 4th ed. Copyright

© 1987 Wm. C. Brown Publishers, Dubuque, Iowa. All Rights Reserved. Reprinted by permission. **Page 87:** From John W. Hole, Jr., *Human Anatomy and Physiology*, 4th ed. Copyright © 1987 Wm. C. Brown Publishers, Dubuque, Iowa. All Rights Reserved. Reprinted by permission. **Page 91:** From Kent M. Van De Graaff and Stuart Ira Fox, *Concepts of Human Anatomy and Physiology.* Copyright © 1986 Wm. C. Brown Publishers, Dubuque, Iowa. All Rights Reserved. Reprinted by permission.

Chapter 13
Page 98: From Kent M. Van De Graaff, *Human Anatomy.* Copyright © 1984 Wm. C. Brown Publishers, Dubuque, Iowa. All Rights Reserved. Reprinted by permission. **Page 99: (top)** From John W. Hole, Jr., *Human Anatomy and Physiology*, 4th ed. Copyright © 1987 Wm. C. Brown Publishers, Dubuque, Iowa. All Rights Reserved. Reprinted by permission. **Page 99: (bottom)** From Kent M. Van De Graaff, *Human Anatomy.* Copyright © 1984 Wm. C. Brown Publishers, Dubuque, Iowa. All Rights Reserved. Reprinted by permission.

Chapter 15
Page 114: From Kent M. Van De Graaff, *Human Anatomy.* Copyright © 1984 Wm. C. Brown Publishers, Dubuque, Iowa. All Rights Reserved. Reprinted by permission. **Page 116:** From John W. Hole, Jr., *Human Anatomy and Physiology*, 4th ed. Copyright © 1987 Wm. C. Brown Publishers, Dubuque, Iowa. All Rights Reserved. Reprinted by permission.

Chapter 16
Page 122: From John W. Hole, Jr., *Human Anatomy and Physiology*, 4th ed. Copyright © 1987 Wm. C. Brown Publishers, Dubuque, Iowa. All Rights Reserved. Reprinted by permission. **Page 123:** From Kent M. Van De Graaff, *Human Anatomy.* Copyright © 1984 Wm. C. Brown Publishers, Dubuque, Iowa. All Rights Reserved. Reprinted by permission.

Chapter 19
Page 141: From John W. Hole, Jr., *Human Anatomy and Physiology*, 4th ed. Copyright © 1987 Wm. C. Brown Publishers, Dubuque, Iowa. All Rights Reserved. Reprinted by permission. **Page 144:** From John W. Hole, Jr., *Human Anatomy and Physiology*, 4th ed. Copyright © 1987 Wm. C. Brown Publishers, Dubuque, Iowa. All Rights Reserved. Reprinted by permission. **Page 145:** From Kent M. Van De Graaff and Stuart Ira Fox, *Concepts of Human Anatomy and Physiology.* Copyright © 1986 Wm. C. Brown Publishers, Dubuque, Iowa. All Rights Reserved. Reprinted by permission.

Chapter 20
Page 151: From John W. Hole, Jr., *Human Anatomy and Physiology*, 2d ed. Copyright © 1981 Wm. C. Brown Publishers, Dubuque, Iowa. All Rights Reserved. Reprinted by permission.